THE MOTION OF A SURFACE BY ITS MEAN CURVATURE

by

Kenneth A. Brakke

Princeton University Press

and

University of Tokyo Press

Princeton, New Jersey

1978

Published in Japan exclusively
by University of Tokyo Press
in other parts of the world by
Princeton University Press

Printed in the United States of America
by Princeton University Press, Princeton, New Jersey

Library of Congress Cataloging in Publication Data Will
be found on the last printed page of this book

TABLE OF CONTENTS

1. Introduction

Surfaces that minimize area subject to various constraints have long been studied. Much of the inspiration for these studies has come from physical systems involving surface tension: soap films, soap bubbles, capillarity, biological cell structure, and others. So far, mathematical investigations have been mostly confined to the equilibrium states of the systems mentioned, with some study of the evolution of non-parametric hypersurfaces [LT]. This work studies in general dimensions a dynamic system: surfaces of no inertial mass drive by surface tension and opposed by frictional force proportional to velocity. The viewpoint is that of geometric measure theory.

The mean curvature vector $\underline{\underline{h}}(V,x)$ of a surface V at a point x can be characterized as the vector which, when multiplied by the surface tension, gives the net force due to surfact tension at that point. For example, if V is a k-sphere with radius R centered at the origin with unit magnitude surface tension, then $\underline{\underline{h}}(V,x) = -kx/|x|^2$. Note that the magnitude of the mean curvature is larger by a factor of k than in many other definitions.

The mathematical object we wish to study may be loosely described as a family of surfaces V_t parameterized by time such that each point at each time is moved with a velocity equal to the mean curvature vector of the surface at that point at that time. A physical system exhibiting this behavior is the motion of grain boundaries in an annealing pure metal. Grain boundaries represent excess energy, and there is effectively a surface tension. It is experimentally observed that these grain boundaries move with a velocity proportional to their mean curvature. For a fuller discussion, see Appendix A.

The question arises: what do we mean by a surface? We do not wish to restrict ourselves to manifolds, firstly because a manifold may evolve singularities, and secondly because systems like grain boundaries are most interesting when they are not manifolds. For surfaces we shall take a certain class of Radon measures known as varifolds, which are defined in section 2.6. The space of varifolds includes anything one would wish to call a surface and has nice compactness properties.

Since it is impossible to follow a measure pointwise, how are we to describe the motion of a surface? We do it by describing how a measure behaves on test functions. Section 2.10 derives an expression for the rate of change of the integral of a test function when the velocity of a surface is a smooth vectorfield. Section 3.3 generalizes this expression to define when a varifold is moving by its mean curvature, even when the mean curvature is far from a smooth vectorfield.

The surface with the simplest nontrivial motion are k-dimensional spheres. Let $R(t)$ denote the radius of a k-sphere at time t. The magnitude of the mean curvature is $k/R(t)$, so $dR(t)/dt = -k/R(t)$. Thus

$$R(t) = (R(0)^2 - 2kt)^{1/2} .$$

The behavior of a k-sphere turns out to be characteristic of the behavior of any k-dimensional surface in the following manner: By 3.7, if a k-dimensional surface at time $t = 0$ is contained in the exterior of a ball of radius $(R(0)^2 - 2kt)^{1/2}$ with the same center. By 3.9, the analogous statement is true of a surface contained in the interior of a ball. An immediate consequence of 3.7 is that a surface starting in a convex set always remains in that convex set.

Section 3.4 establishes bounds on the motion of a surface moving by its mean curvature, and these bounds are used in 3.10 to show that the motion is continuous except for instantaneous losses of area.

Chapter 4 addresses the problem of the existence of a surface moving by its mean curvature with a given initial surface. We consider the k-dimensional initial surfaces to be members of a class of rectifiable varifolds with a positive lower bound on the k-dimensional densities. This class includes all surfaces of interest. For such an initial surface, Chapter 4 gives an approximation procedure that yields a one parameter family of varifolds that satisfies the definition of motion by mean curvature given in 3.3 and are rectifiable at almost all times. If the initial surface has integer densities, which all familiar surfaces do, then the constructed varifolds are also integral at almost all times. These properties are proven in 4.29.

If the initial surface were a smooth manifold, then one might get a solution for a short time by the straightforward use of the theory of partial differential equations, as briefly discussed in 3.1. My procedure will yield the same result in such a case, as noted in 4.15.

For a given initial surface, the subsequent motion may be naturally not unique, as illustrated in C.4. Therefore my procedure does not strive for uniqueness.

Certain modifications may easily be made at one stage to model different types of behavior (see Remark 2 of 4.9). None of these modifications affect the results of Chapter 4.

Chapter 5 proves that the mean curvature vector is almost everywhere perpendicular to an integral varifold whenever the notion of mean curvature vector is valid. This perpendicularity is easily proved for class 2 manifolds in differential geometry, but under our very broad hypotheses, we will have to delve deep into the microstructure of varifolds. This result is not directly concerned with moving varifolds, but it is essential for chapter 6.

Chapter 6 shows that a unit density integral varifold moving by its mean curvature is an infinitely differentiable manifold almost everywhere, except perhaps when there is an instantaneous loss of area. "Unit density" means that the density of the varifold is the same everywhere. Without this hypothesis, not even minimal varifolds (those with zero mean curvature) are known to be regular. Section 6.2 describes an example illustrating the problems that arise with multiple densities. Unfortunately, the existence construction of chapter 4 has not yet been made to yield unit density varifolds. However, the regularity proof is totally independent of the source of a varifold moving by its mean curvature, and would apply, for example, to the non-parametric hypersurfaces of [LT].

Appendix A discusses metal grain boundaries, as mentioned above. Appendix B discusses smooth simple closed curves in the plane moving by their mean curvature. Appendix C discusses 1-dimensional surfaces in a plane that retain the same shape but change in size as they move by mean curvature. There are computer plots of several such surfaces. Appendix D shows the necessity of the hypothesis of positive lower bounds on densities in Chapter 4 by describing a rectifiable varifold with densities approaching zero that should turn unrectifiable as it moves by its mean curvature.

As for generalizations of this work, everything would still be valid on smooth Riemannian manifolds, when properly interpreted. Extension ot integrands much different from the area integrand would not be as easy, because Allard [AW2] has shown that essentially only the area integrand satisfies monotonicity theorems, such as 4.17, which are vital to the methods herein.

I thank my advisor, Professor Frederick J. Almgren, Jr., for his guidance, for many inspiring discussions, and especially for his never-ending optimism and encouragement. I am grateful to the National Science Foundation for support.

2. Preliminaries

2.1 General definitions.

We follow the standard terminology of [FH]. Most of the definitions regarding varifolds come from [AW1].

We denote by $\mathbb{N}$ the positive integers and by $\mathbb{R}$ the real numbers. Throughout this paper k and n are fixed positive integers with $k \le n$. Define

$$\mathbb{R}^+ = \{t \in \mathbb{R}: t \ge 0\} ,$$

$$\mathbb{U}^k(a,r) = \{x \in \mathbb{R}^k: |x - a| < r\} ,$$

$$\mathbb{U}(a,r) = \{x \in \mathbb{R}^n: |x - a| < r\} ,$$

$$\mathbb{B}^k(a,r) = \{x \in \mathbb{R}^k: |x - a| < r\},$$

$$\mathbb{B}(a,r) = \{x \in \mathbb{R}^n: |x - a| < r\}.$$

Frequently, we will treat $\mathbb{R}^k$ as a subspace of $\mathbb{R}^n$.

We will use $\int dx$ to denote integration with respect to Lebesgue measure $\mathscr{L}^n$ on $\mathbb{R}^n$. Set

$$\underline{\underline{\alpha}} = \mathscr{L}^k\mathbb{B}^k(0,1).$$

We denote by $\mathscr{H}^k$ Hausdorff k-dimensional measure on $\mathbb{R}^n$.

We will often use $\langle f,g \rangle$ to denote the value of a distribution f on an appropriate test function g, especially when dealing with convolutions.

For $\psi \in \underline{\underline{C}}_0^2(\underline{\underline{R}}^n, \underline{\underline{R}}^+)$ we define

$$(|D\psi|^2/\psi)(x) = \begin{cases} |D\psi(x)|^2/\psi(x) & \text{if } \psi(x) \neq 0 \\ 0 & \text{if } \psi(x) = 0. \end{cases}$$

It can be shown that $|D\psi|^2/\psi$ is bounded.

If $F:\underline{\underline{R}} \to \underline{\underline{R}}$ then we define for each $t \in \underline{\underline{R}}$ the upper derivate of F at t by

$$\bar{D}F(t) = \limsup_{s \to t} \frac{F(s) - F(t)}{s - t} .$$

We shall also use the upper left and upper right derivates, denoted D^- and D^+ respectively.

2.2 The Besicovitch covering theorem.

There is a positive integer $\underline{\underline{B}}(n)$ with the following property: If A is a subset of $\underline{\underline{R}}^n$ and C is a family of closed balls in $\underline{\underline{R}}^n$ such that each point of A is the center of a member of C, then there are disjoint subfamilies C_i, $i = 1,\ldots, \underline{\underline{B}}(n)$, of C such that

$$A \subset \cup\{\cup C_i: i = 1,\ldots,\underline{\underline{B}}(n)\} .$$

For the proof, see [FH 2.8.14].

2.3. Homothety and translation.

For each $r \in R$ we define the homothety $\underline{\underline{\mu}}(r):\underline{\underline{R}}^n \to \underline{\underline{R}}^n$ by

$$\underline{\underline{\mu}}(r)(x) = rx.$$

For each $a \in \underline{\underline{R}}^n$ we define the translation $\underline{\underline{\tau}}(a):\underline{\underline{R}}^n \to \underline{\underline{R}}^n$ by

$$\underline{\underline{\tau}}(a)(x) = x - a.$$

2.4. Densities and tangent cones.

If μ is a measure on $\underline{\underline{R}}^n$ then we define the k-dimensional upper density and density of μ at $a \in \underline{\underline{R}}^n$ by

$$\Theta^{*k}(\mu,a) = \limsup_{r \to 0^+} \mu\underline{\underline{B}}(a,r)/\underline{\underline{\alpha}}r^k,$$

$$\Theta^k(\mu,a) = \lim_{r \to 0^+} \mu\underline{\underline{B}}(a,r)/\underline{\underline{\alpha}}r^k .$$

The approximate tangent cone of μ at $a \in \underline{\underline{R}}^n$ is $\mathrm{Tan}^k(\mu,a) = \bigcap\{C:$ C is a cone in $\underline{\underline{R}}^n$ with vertex at a and

$$\Theta^k(\mu | \underline{\underline{R}}^n \sim C,\ a) = 0\} .$$

2.5. The grassman manifold, homomorphisms, and cylinders.

Let $\underline{\underline{G}}(n,k)$ denote the space of k-dimensional subspaces of $\underline{\underline{R}}^n$, which can also be thought of as the set of unit simple k-vectors. Suppose $S \in \underline{\underline{G}}(n,k)$. We will also use S to denote orthogonal projection of $\underline{\underline{R}}^n$ onto S.

Let $\underline{\underline{G}}_k(\underline{\underline{R}}^n) = \underline{\underline{R}}^n \times \underline{\underline{G}}(n,k)$.

For $A, B \in \mathrm{Hom}(\underline{\underline{R}}^n,\underline{\underline{R}}^n)$, we define a scalar product $A \cdot B$ by

$$A \cdot B = \mathrm{trace}(A^* \circ B).$$

The identity element of $\operatorname{Hom}(\underline{\underline{R}}^n, \underline{\underline{R}}^n)$ will be denoted by $\underline{\underline{I}}$. If $\phi \in \underline{\underline{C}}^2(\underline{\underline{R}}^n, \underline{\underline{R}})$ or $g \in \underline{\underline{C}}^1(\underline{\underline{R}}^n, R^n)$, then we will sometimes treat $D\phi(x)$ as an element of $\underline{\underline{R}}^n$ and $D^2\phi(x)$ or $Dg(x)$ as elements of $\operatorname{Hom}(\underline{\underline{R}}^n, \underline{\underline{R}}^n)$. The tensor product $v \otimes w$ of two vectors $v, w, \in \underline{\underline{R}}^n$ is also in $\operatorname{Hom}(\underline{\underline{R}}^n, \underline{\underline{R}}^n)$. The norm $\| \;\; \|$ on $\operatorname{Hom}(\underline{\underline{R}}^n, \underline{\underline{R}}^n)$ is

$$\|A\| = \sup\{|A(x)| : x \in R^n, \ |x| = 1\}.$$

We will frequently use the following facts about $S, T \in \underline{\underline{G}}(n,k)$ and A, g, ϕ, v, w as above:

$$v \otimes w \cdot S = S(v) \cdot w = v \cdot S(w) = S(v) \cdot S(w),$$

$$D^2\phi(x)(v,w) = v \otimes w \cdot D^2\phi(x),$$

$$\underline{\underline{I}} \cdot S = k, \quad 0 \le k - S \cdot T \le k\|S - T\|^2$$

$$|A \cdot S| \le k\|A\|,$$

$$|T(S^{\perp}(w))| \le \|S - T\| \, |w|, \text{ and}$$

$$|T(S^{\perp}(T(w)))| \le \|S - T\|^2 |w|.$$

For $T \in \underline{\underline{G}}(n,k)$, $a \in \underline{\underline{R}}^n$, and $0 < r < \infty$, we define the cylinder

$$\underline{\underline{C}}(T,a,r) = \{x \in R^n : |T(x-a)| \le r\}.$$

2.6 Varifolds.

We say V is a k-dimensional varifold in $\mathbb{R}^n$ if and only if V is a Radon measure on $\mathbb{G}_k(\mathbb{R}^n)$. Let $\mathbb{V}_k(\mathbb{R}^n)$ be the weakly topologized space of k-dimensional varifolds in $\mathbb{R}^n$. Whenever $V \in \mathbb{V}_k(\mathbb{R}^n)$, we define the weight of V to be the Radon measure $\|V\|$ on $\mathbb{R}^n$ given by

$$\|V\|A = V\{(x,S) \in \mathbb{G}_k(\mathbb{R}^n): x \in A\}$$

whenever A is a Borel subset of $\mathbb{R}^n$. We let $V^{(\cdot)}$ be the $\|V\|$ measurable function with values in the Radon measures on $\mathbb{G}(n,k)$ such that for any $\psi \in \mathbb{C}_0(\mathbb{G}_k(\mathbb{R}^n),\mathbb{R})$

$$V(\psi) = \int\int \psi(x,S)dV^{(x)}S \, d\|V\|x.$$

If $\beta:\mathbb{G}_k(\mathbb{R}^n) \to \mathbb{R}$ is a locally V summable function, then $V \llcorner \beta \in \mathbb{V}_k(\mathbb{R}^n)$ is defined by

$$(V \llcorner \beta)A = \int_A \beta(x,S)dV(x,S).$$

The same notation will be used with the obvious meaning even if β is only defined on $\mathbb{R}^n$. Similarly, if A is a $\|V\|$ measurable subset of $\mathbb{R}^n$, we will use $V \llcorner A$ to abbreviate the restriction $V \llcorner [A \times \mathbb{G}(n,k)]$, and $\int_A \beta(x,S)dV(x,S)$ to abbreviate $\int_{A \times \mathbb{G}(n,k)} \beta(x,S)dV(x,S)$.

By the well-known compactness properties of Radon measures, the set of varifolds

$$\{V \in \mathbf{V}_k(\mathbf{R}^n): \ \|V\|\mathbf{B}(0,R_i) \le B_i, \quad i \in \mathbf{N}\}$$

is compact if $B_i < \infty$ for all i and $\lim_{i \to \infty} R_i = \infty$.

2.7. Rectifiable and integral varifolds.

Whenever E is an $\mathcal{H}^k$ measurable subset of $\mathbf{R}^n$ which meets every compact subset of $\mathbf{R}^n$ in an $(\mathcal{H}^k, k)$ rectifiable subset [FH 3.2.14], there is a naturally associated varifold $\mathbf{v}(E) \in \mathbf{V}_k(\mathbf{R}^n)$ defined by

$$\mathbf{v}(E)A = \mathcal{H}^k\{x: (x, \operatorname{Tan}^k(\mathcal{H}^k \llcorner E, x)) \in A\}$$

whenever $A \subset \mathbf{G}_k(\mathbf{R}^n)$.

We say a varifold $V \in \mathbf{V}_k(\mathbf{R}^n)$ is a k-dimensional rectifiable varifold if there are positive real numbers $c_1, c_2, \ldots$ and $\mathcal{H}^k$ measurable subsets $E_1, E_2, \ldots$ of $\mathbf{R}^n$ which meet every compact subset of $\mathbf{R}^n$ in an $(\mathcal{H}^k, k)$ rectifiable subset such that

$$V = \sum_{i=1}^{\infty} c_i \mathbf{v}(E_i) .$$

If the c_i may be taken to be positive integers, then we say V is a k-dimensional integral varifold. We let

$$\mathbf{RV}_k(\mathbf{R}^n) \quad \text{and} \quad \mathbf{IV}_k(\mathbf{R}^n)$$

be the spaces of k-dimensional rectifiable and integral varifolds respectively.

2.8. Mapping of varifolds. [AF I.1(13)]

Suppose $f:\mathbb{R}^n \to \mathbb{R}^n$ is a proper mapping of class 1 and $V \in \mathbf{V}_k(\mathbb{R}^n)$. Then the varifold $f_\# V \in \mathbf{V}_k(\mathbb{R}^n)$ induced by f is characterized by the condition

$$(f_\# V)A = \int_{\{(x,T):[f(x),Df(x)(T)]\in A\}} |\Lambda_k(Df(x)\circ T)|\,dV(x,T)$$

whenever A is a Borel subset of $\mathbf{G}_k(\mathbb{R}^n)$.

Suppose $f:\mathbb{R}^n \to \mathbb{R}^n$ is a proper Lipschitz map and $V \in \mathbf{RV}_k(\mathbb{R}^n)$. Then the induced varifold $f_\# V \in \mathbf{RV}_k(\mathbb{R}^n)$ is characterized by

$$(f_\# V)A = \int_{\{(x,T):[f(x),\operatorname{apDf}(x)(T)]\in A\}} |\Lambda_k(\operatorname{apDf}(x)\circ T)|\,dV(x,T)$$

whenever A is a Borel subset of $\mathbf{G}_k(\mathbb{R}^n)$; here the approximate differential is

$$\operatorname{apDf}(x) = (\|V\|,k)\operatorname{apDf}(x) : \operatorname{Tan}^k(\|V\|,x) \to \mathbb{R}^n ,$$

see [FH 3.2.16, 3.2.19, 3.2.20]. The function $f_\#: \mathbf{RV}_k(\mathbb{R}^n) \to \mathbf{RV}_k(\mathbb{R}^n)$ is not in general continuous. One observes $f_\#(\mathbf{IV}_k(\mathbb{R}^n)) \subset \mathbf{IV}_k(\mathbb{R}^n)$.

2.9. First variation.

Suppose $\varepsilon > 0$, $h:(-\varepsilon,\varepsilon) \times \mathbb{R}^n \to \mathbb{R}^n$ is smooth, $h_t(x) = h(t,x)$ for $(t,x) \in (-\varepsilon,\varepsilon) \times \mathbb{R}^n$, $h_0(x) = x$, and the set

$$\{x: h_t(x) \neq x \quad \text{for some} \quad t \in (-\varepsilon, \varepsilon)\}$$

has compact closure in an open subset G of $\underline{\underline{R}}^n$. Let

$$g = (\partial h/\partial t)(0,\cdot) \in \underline{\underline{C}}_0^1(\underline{\underline{R}}^n, \underline{\underline{R}}^n).$$

Then for $V \in \underline{\underline{V}}_k(\underline{\underline{R}}^n)$ such that $\|V\|G < \infty$ we have by [AW1 4.1]

$$(d/dt)\|h_{t\#}V\|G\Big|_{t=0} = \int Dg(x)\cdot S dV(x,S).$$

This motivates for any $V \in \underline{\underline{V}}_k(\underline{\underline{R}}^n)$ the definition of a linear function

$$\delta V: \underline{\underline{C}}_0^1(\underline{\underline{R}}^n, \underline{\underline{R}}^n) \to \underline{\underline{R}},$$

called the <u>first variation</u> of V, by

$$\delta V(g) = \int Dg(x)\cdot S \, dV(x,S).$$

If $\delta V = 0$, then V is called <u>stationary</u>. We define the <u>total variation</u> $\|\delta V\|$ to be the largest Borel regular measure on $\underline{\underline{R}}^n$ determined by

$$\|\delta V\|G = \sup\{\delta V(g): g \in \underline{\underline{C}}_0^1(\underline{\underline{R}}^n, \underline{\underline{R}}^n), \text{ spt } g \subset G \text{ and } |g| \leq 1\}$$

whenever G is an open subset of $\underline{\underline{R}}^n$.

If $\|\delta V\|$ is a Radon measure, then there is a $\|\delta V\|$ measurable function $\eta(V;\cdot)$ with values in $\underline{S}^{n-1}$ such that

$$\delta V(g) = \int g(x)\cdot \eta(V;x) d\|\delta V\|x$$

for $g \in \mathbf{C}_0^1(R^n, R^n)$. The theory of symmetrical derivation (see [FH 2.8.18, 2.9]) implies the following: The formula

$$\|\delta V\| / \|V\|(x) = \lim_{r \downarrow 0} \|\delta V\| \mathbf{B}(x,r) / \|V\| \mathbf{B}(x,r)$$

defines a real-valued $\|V\|$ measurable function on R^n such that if

$$\|\delta V\|_{sing} = \|\delta V\| \llcorner \{x: \|\delta V\|/\|V\|(x) = \infty\}$$

then

$$\|\delta V\| B = \int_B \|\delta V\|/\|V\|(x) \, d\|V\|x + \|\delta V\|_{sing} B$$

whenever B is a Borel subset of $\mathbf{R}^n$. The formula

$$\mathbf{h}(V,x) = -\|\delta V\|/\|V\|(x) \eta(V;x)$$

defines a $\|V\|$ measurable function with values in $\mathbf{R}^n$ such that

$$\delta V(g) = -\int g(x) \cdot \mathbf{h}(V,x) \, d\|V\|x + \int g(x) \cdot \eta(V;x) \, d\|\delta V\|_{sing} x$$

whenever g is a Borel measurable function with values in $\mathbf{R}^n$ such that $\int |g(x)| \, d\|\delta V\| x < \infty$. If $\|\delta V\|_{sing} = 0$, then we call $\mathbf{h}(V, \cdot)$ the generalized mean curvature vector of V.

The preceding mathematics has a physical interpretation. A surface naturally corresponds to a varifold. When a surface has a surface tension, the area is proportional to the total energy. If g is the velocity of the surface, then the rate of change of energy (the power) is $\delta V(g)$. Since power is integral of the force times the velocity over the surface, clearly $\underline{h}(V,x)$ is proportional to the force due to the surface tension. Singular first variation, $\|\delta V\|_{sing}$, occurs at edges, sharp corners, and the like.

2.10 First variation with respect to other integrands.

Suppose $\phi \in \underline{C}_0^2(\underline{R}^n,\underline{R})$ and V, G, ε, h, and g are as in 2.9. Then by [AW1 4.9(1)]

$$(d/dt)\,\|h_{t\#}V\|(\phi)\Big|_{t=0} = \int Dg(x)\cdot S\;\phi(x)\,dV(x,S) + \int g(x)\cdot D\phi(x)\,d\|V\|x.$$

We are led to define the first variation $\delta(V,\phi)$ of V with respect to ϕ by setting

$$(1)\quad \delta(V,\phi)(g) = \int Dg(x)\cdot S\phi(x)\,dV(x,S) + \int g(x)\cdot D\phi(x)\,d\|V\|x$$

whenever $g \in \underline{C}_0^1(\underline{R}^n,\underline{R}^n)$.

Proposition: If $V \in \underline{V}_k(\underline{R}^n)$, $\phi \in \underline{C}^1(\underline{R}^n,\underline{R})$ and $g \in \underline{C}_0^1(\underline{R}^n,\underline{R}^n)$, then

$$(2)\quad \delta(V,\phi)(g) = \delta V(\phi g) - \int S(D\phi(x))\cdot g(x)\,dV(x,S) + \int D\phi(x)\cdot g(x)\,d\|V\|x.$$

Proof: From (1)

$$\delta(V,\phi)(g) = \int Dg(x)\cdot S\phi(x)\,dV(x,S) + \int D\phi(x)\cdot g(x)\,d\|V\|x$$

$$= \int D(\phi g)(x)\cdot S - (D\phi(x) \otimes g(x))\cdot S\,dV(x,S)$$

$$+ \int D\phi(x)\cdot g(x)\,d\|V\|x$$

$$= \delta V(\phi g) - \int S(D\phi(x))\cdot g(x)\,dV(x,S)$$

$$+ \int D\phi(x)\cdot g(x)\,d\|V\|x. \qquad \square$$

Note that we may also write (2) as

(3) $$\delta(V,\phi)(g) = \delta V(\phi g) + \int S^{\perp}(D\phi(x))\cdot g(x)\,dV(x,S),$$

or if $\|\delta V\|$ is a Radon measure and $\|\delta V\|_{sing} = 0$, then

(4) $$\delta(V,\phi)(g) = \int \underline{h}(V,x)\cdot g(x)\phi(x)\,d\,V\,x + \int S^{\perp}(D\phi(x))\cdot g(x)\,dV(x,S).$$

2.11. Compactness theorem for rectifiable varifolds.

Theorem [AW1 5.6]: Suppose $G_1, G_2, \ldots$ are open subsets of $\mathbb{R}^n$, $\mathbb{R}^n = \bigcup_{i=1}^{\infty} G_i$, $M_1, M_2, \ldots$ are nonnegative real numbers, and θ is a positive real valued continuous function on $\mathbb{R}^n$. The set of those varifolds V in $\mathbf{RV}_k(\mathbb{R}^n)$ for which

$$(\|V\| + \|\delta V\|)G_i \le M_i, \quad i = 1,2,\ldots,$$

and

$$\Theta^k(\|V\|,x) \geq \theta(x) \quad \text{for} \quad \|V\| \quad \text{almost all} \quad x \in \mathbb{R}^n$$

is compact. □

2.12. Compactness theorem for integral varifolds.

Theorem [AW1 6.4]: Suppose $G_1, G_2, \cdots$ are open subsets of $\mathbb{R}^n$, $\mathbb{R}^n = \bigcup_{i=1}^{\infty} G_i$, and $M_1, M_2, \ldots$ are nonnegative real numbers. Then the set of those varifolds V in $\mathbf{IV}_k(\mathbb{R}^n)$ for which

$$(\|V\| + \|\delta V\|)G_i \leq M_i, \quad i = 1,2,\ldots$$

is compact. □

3. Motion by mean curvature

3.1. Manifold difficulties.

On first considering the problem of a surface moving by its mean curvature, one is likely to try to apply results from the theory of partial differential equations. In what is called the parametric approach, the moving surface is viewed as a family of maps $F_t: \mathbb{R}^k \to \mathbb{R}^n$. From differential geometry [SM, p. 193], the mean curvature vector $h_t(x)$ at $F_t(x)$ is the invariant Laplacian of the position vector:

(1) $$h_t(x) = \Delta F_t(x).$$

In coordinates, this is

(2) $$h_t(x)_m = \sum_{i,j=1}^{k} \frac{1}{g(x)} \frac{\partial}{\partial x_i} [g(x) g^{ij}(x) \frac{\partial F_t(x)_m}{\partial x_j}], \quad m = 1,\cdots,k,$$

where (g^{ij}) is the inverse matrix of the metric (g_{ij}),

$$g_{ij}(x) = \sum_{p=1}^{n} \frac{\partial F_t(x)_p}{\partial x_i} \frac{\partial F_t(x)_p}{\partial x_j},$$

and $g^2 = |\det(g_{ij})|$. Thus, the problem becomes to solve

(3) $$\partial F_t(x)/\partial t = \Delta F_t(x).$$

This looks like a vector-valued heat equation, except that the operator Δ depends on F_t.

Equation (3) is parabolic, just as the minimal surface equation $\Delta F(x) = 0$ is well known to be elliptic. The theory of systems of quasilinear parabolic partial differential equations applies. For example, if F_0 is nice enough, then [ES III.4] guarantees the existence of F_t for some short time interval.

The non-parametric approach is to represent a moving surface as the graph of maps $f_t\colon \mathbb{R}^k \to \mathbb{R}^{n-k}$. Here, the equation of motion becomes

$$(4) \qquad \partial f_t(x)/\partial t = \sum_{i,j=1}^{k} g^{ij}(x)\,\partial^2 f_t(x)/\partial x_i \partial x_j,$$

where the metric arises from $F_t = \mathbb{I} \oplus f_t$. This equation is also nicely parabolic, and it is nearly the heat equation when f_t is nearly constant. The analogy to heat will be a guiding principle in the regularity theory of chapter 6. There we will also use the fact that solutions to (4) are infinitely differentiable [ES II.1.5].

There are many objections to these two approaches. The principal one is the topological restriction placed on surfaces. Real grain boundaries are full of singularities, and the topological type continually changes. Even if the initial surface is representable parametrically, the existence of a solution is guaranteed only for a short time, as the surface may develop knots and other singularities. The non-parametric problem may have "generalized solutions" [LT] existing forever, but puts even more drastic restrictions on the type of surface.

The varifold approach places no restrictions on the nature of a surface. Anything with area and tangent planes is a varifold. Of course, that means (3) or (4) no longer apply. Therefore, the first task of this chapter is to provide a definition of motion by mean curvature for varifolds that can always be applied. The starting point for this definition is the first variation with respect to an integrand, discussed in 2.10. We see from 2.10 that if a varifold V_t represents a smooth manifold, then (3) is equivalent to requiring

$$(5) \qquad (d/dt)\|V_t\|(\phi) = \delta(V_t,\phi)(\underline{\underline{h}}(V_t,\cdot))$$

for smooth test functions ϕ. We will generalize (5) to all V, but first we must define $\delta(V,\phi)(\underline{\underline{h}}(V,\cdot))$ for all $V \in \underline{\underline{V}}_k(\underline{\underline{R}}^n)$.

3.2. <u>Definition of</u> $\delta(V,\phi)\underline{\underline{h}}(V,\cdot))$.

Suppose $V \in \underline{\underline{V}}_k(\underline{\underline{R}}^n)$ and $\phi \in \underline{\underline{C}}_0^1(R^n,R^+)$. If $\|\delta V\| \llcorner \phi$ is not a Radon measure, if $\|\delta V\|_{sing} \llcorner \phi = 0$, or if

$$(1) \qquad \int |\underline{\underline{h}}(V,x)|^2 \phi(x)\, d\|V\|x = \infty,$$

then we will set

$$(2) \qquad \delta(V,\phi)(\underline{\underline{h}}(V,\cdot)) = -\infty.$$

Otherwise, in analogy with 2.10 (4), set

(3) $$\delta(V,\phi)(h(V,\cdot)) = -\int |\underline{\underline{h}}(V,x)|^2\phi(x)\,d\|V\|x + \int S^\perp(D\phi(x))\cdot\underline{\underline{h}}(V,x)\,dV(x,S).$$

<u>Remarks</u>: To enable us to write single formulas to cover all cases, we will make the convention that

$$\int |\underline{\underline{h}}(V,x)|^2\phi(x)\,d\|V\|x = \infty$$

also in case $\|\delta V\| \llcorner \phi$ is not a Radon measure or $\|\delta V\|_{sing} \llcorner \phi \neq 0$. This makes (2) formally consistent with (3).

Since $\underline{\underline{h}}(V,\cdot)$ may not be bounded, even on compact sets, it is not clear <u>a priori</u> that the rate of change of $\|V_t\|(\phi)$ should be given by $\delta(V,\phi)(\underline{\underline{h}}(V,\cdot))$. However, we shall see in 3.4 that unbounded mean curvature does not lead to unbounded rates of growth on test functions.

3.3. <u>Varifold moving by its mean curvature</u>.

We shall say that a one parameter family of varifolds $V_t \in \underline{\underline{V}}_k(R^n)$, $t \in R^+$, is a <u>varifold moving by its mean curvature</u> if and only if

(1) $$\overline{D}\|V_t\|(\phi) \leq \delta(V_t,\phi)(\underline{\underline{h}}(V_t,\cdot))$$

for every $\phi \in \underline{\underline{C}}_0^1(\underline{\underline{R}}^n,\underline{\underline{R}}^+)$ and for all $t \in \underline{\underline{R}}^+$.

<u>Remarks</u>: The notion of derivate is used because $\|V_t\|(\phi)$ may not always be differentiable, or even continuous (see

Appendix C.5), and the upper derivate gives a stronger condition than any other derivate. We will see in 3.10(b) that (1) implies $\|V_t\|(\phi)$ is differentiable for almost all $t \in \underline{\underline{R}}^+$, but it is not clear whether we should require equality in (1) for almost every t. Appendix C.4 shows an example in which V_0 has zero mean curvature, yet we want $\overline{D}\|V_t\|(\phi)\,|_{t=0} = -\infty$ if $\phi(0) > 0$. It is conceivable that there is some example in which frequent behavior of this sort leads to

$$\overline{D}\|V_t\|(\phi) < \delta(V_t,\phi)(\underline{\underline{h}}(V_t,\cdot))$$

for all $t \in \underline{\underline{R}}^+$. Condition (1) is also the condition that naturally arises out of the construction of Chapter 4, as remarked in 4.18.

This definition does not imply anything about the uniqueness of a varifold moving by its mean curvature for a given initial varifold. Appendix C.4 gives one example of non-uniqueness.

For general varifolds, 3.1(2) cannot completely characterize the motion because it says nothing about the rate of change of the Grassman manifold component $V_t^{(\cdot)}$ of V_t (see 2.6). It would obviously be nice to use first variation with respect to a test function defined on $\underline{\underline{G}}_k(\underline{\underline{R}}^n)$, but such a first variation could not be converted into a form like 2.10(1) which could be generalized from smooth vectorfields g to mean curvature $\underline{\underline{h}}(V,x)$ as in 3.2(3). However, for rectifiable varifolds $\|V\|$ does determine V, and this covers almost all interesting cases.

This chapter henceforth will deal only with consequences of (1). Existence of V_t for certain V_0 will be shown in Chapter 4.

3.4. Upper bound on motion.

Proposition: If $V \in \mathbf{V}_k(\mathbf{R}^n)$ and $\phi \in \mathbf{C}_0^2(\mathbf{R}^n, \mathbf{R}^+)$ then

$$\delta(V,\phi)(\mathbf{h}(V,\cdot)) \le - \int |\mathbf{h}(V,x)|^2 \phi(x)\, d\|V\|x$$

$$+ [\int |\mathbf{h}(V,x)|^2 \phi(x)\, d\|V\|x]^{1/2} \|V\|(|D\phi|^2/\phi)^{1/2}$$

$$\le \|V\|(|D\phi|^2/\phi).$$

Proof: If $\delta(V,\phi)(\mathbf{h}(V,\cdot)) = -\infty$ then we are done. Otherwise, by 3.2(3)

$$\delta(V,\phi)(\mathbf{h}(V,\cdot)) = - \int |\mathbf{h}(V,x)|^2 \phi(x)\, d\|V\|x$$

$$+ \int S^{\perp}(D\phi(x)) \cdot \mathbf{h}(V,x)\, dV(x,S)$$

$$\le - \int |\mathbf{h}(V,x)|^2 \phi(x)\, d\|V\|x + \int |D\phi(x)|\, |\mathbf{h}(V,x)|\, d\|V\|x,$$

from which the conclusions follow by applying the Schwarz inequality to the second term on the right hand side and finding the maximum value of the resulting expression. □

Remark: This shows $\delta(V,\phi)(\mathbf{h}(V,\cdot)) \to -\infty$ as $\int |\mathbf{h}(V,x)|^2 \phi(x)\, d\|V\|x \to \infty$ for a bounded value of $\|V\|(|D\phi|^2/\phi)$, justifying definition 3.2(2).

3.5. Time varying test functions.

Proposition: If V_t is a varifold moving by its mean curvature, $0 \leq r < s < \infty$, and $\psi \in \underline{\underline{C}}_0^1([r,s] \times \underline{\underline{R}}^n, \underline{\underline{R}}^+)$, then

$$D^+ \|V_t\|(\psi(t,\cdot)) \leq \delta(V_t,\psi(t,\cdot))(\underline{\underline{h}}(V_t,\cdot)) + \|V_t\|(\partial\psi(t,\cdot)/\partial t)$$

for $t \in [t,s)$.

Proof: Let $\psi \in \underline{\underline{C}}_0^2(\underline{\underline{R}}^n,\underline{\underline{R}}^+)$ be such that $\psi(x) \geq 1$ if $x \in \operatorname{spt} \psi(t,\cdot)$ for any $t \in [r,s]$.

Suppose $t \in [r,s)$. It follows from 3.3(1) and 3.4 that there are $M < \infty$ and $\delta > 0$ such that $\|V_u\|(\psi) < M$ for $t \leq u \leq t + \delta$. We may write

$$D^+ \|V_t\|(\psi(t,\cdot))$$

$$= \limsup_{\Delta t \downarrow 0} [\|V_{t+\Delta t}\|(\psi(t+\Delta t,\cdot)) - \|V_t\|(\psi(t,\cdot))]/\Delta t$$

$$\leq \limsup_{\Delta t \downarrow 0} [\|V_{t+\Delta t}\|(\psi(t,\cdot)) - \|V_t\|(\psi(t,\cdot))]/\Delta t$$

$$+ \limsup_{\Delta t \downarrow 0} \|V_{t+\Delta t}\|(\partial\psi(t,\cdot)/\partial t)$$

$$+ \limsup_{\Delta t \downarrow 0} (1/\Delta t) \iint_0^{\Delta t} |\partial\psi(t+\theta,x)/\partial t - \partial\psi(t,x)/\partial t| \, d\theta \, d\|V_{t+\Delta t}\| x.$$

By the definition of motion by mean curvature,

$$\limsup_{\Delta t \downarrow 0} [\|V_{t+\Delta t}\|(\psi(t,\cdot)) - \|V_t\|(\psi(t,\cdot))]/\Delta t$$

$$\leq \delta(V_t, \psi(t,\cdot))(\underline{h}(V_t,\cdot)).$$

By approximating $\partial\psi(t,\cdot)/\partial t$ with class 2 test functions, we see from 3.4 that

$$\limsup_{\Delta t \downarrow 0} \|V_{t+\Delta t}\|(\partial\psi(t,\cdot)/\partial t) \leq \|V_t\|(\partial\psi(t,\cdot)/\partial t).$$

Finally, by the continuity of $\partial\psi/\partial t$, compactness, and the boundedness of $\|V_{t+\Delta t}\|(\phi)$ for $\Delta t < \delta$,

$$\limsup_{\Delta t \downarrow 0} (1/\Delta t) \int\int_0^{\Delta t} |\partial\psi(t+\theta,x)/\partial t - \partial\psi(t,x)/\partial t| \, d\theta \, d\|V_{t+\Delta t}\| x$$

$$\leq \limsup_{\Delta t \downarrow 0} [M \sup\{|\partial\psi(t+\theta,x)/\partial t - \partial\psi(t,x)/\partial t| : x \in \underline{\underline{R}}^n,\ 0 \leq \theta \leq \Delta t\}]$$

$$= 0. \qquad \square$$

Remark: The proposition is also true for D^-, but 3.6 is needed first to provide an upper bound for $\|V_{t+\Delta t}\|(\phi)$ for $\Delta t < 0$. However, D^+ will be sufficient for all our needs.

3.6. Barrier functions.

A class 2 function $\Psi: \underline{\underline{R}}^+ \times \underline{\underline{R}}^n \to \underline{\underline{R}}^+$ will be called a <u>barrier function</u> if there exist $\phi \in \underline{\underline{C}}^2(\underline{\underline{R}}, \underline{\underline{R}}^+)$ and $a \in \underline{\underline{R}}^n$ such that

$$\Psi(t,x) = \phi(|x-a|^2 + 2kt)$$

for all $(t,x) \in \mathbf{R}^+ \times \mathbf{R}^n$ and

$$(d\phi(r)/dr)^2 \leq 4\phi(r)d^2\phi(r)/dr^2$$

for all $r \in \mathbf{R}$.

Theorem: *If V_t is a varifold moving by its mean curvature and Ψ is a barrier function with compact support, then*

$$D^+\|V_t\|(\Psi(t,\cdot)) \leq 0$$

for $t \in \mathbf{R}^+$.

Proof: Let $t \in \mathbf{R}^+$. If $\delta(V_t,\Psi(t,\cdot))(\underline{h}(V_t,\cdot)) = -\infty$ then we are done. Otherwise, letting $V = V_t$, we may rewrite 3.2(3) as

$$(1) \qquad \delta(V,\Psi(t,\cdot))(\underline{h}(V,x)) = \int -|\underline{h}(V,x)|^2\Psi(t,x)$$

$$- \underline{h}(V,x)\cdot S(D_x\Psi(t,x))dV(x,S) + \int \underline{h}(V,x)\cdot D_x\Psi(t,x)d\|V\|x.$$

Completing the square in the first integral, noting that $D\Psi(t,x) = 0$ when $\Psi(t,x) = 0$, and using

$$\int \underline{h}(V,x)\cdot D_x\Psi(t,x)d\|V\|x = -\delta V(D_x\Psi(t,\cdot))$$

$$= -\int D^2\Psi(t,x)\cdot SdV(x,S)$$

gives

$$(2) \qquad \delta(V,\Psi(t,\cdot))(\underline{\underline{h}}(V,x))$$

$$\leq \int_{\{(x,S);\Psi(t,x)>0\}} -|\underline{\underline{h}}(V,x)\Psi(t,x)^{1/2}+(1/2)S(D_x\Psi(t,x))\Psi(t,x)^{-1/2}|^2$$

$$+ (1/4)|S(D_x\Psi(t,x))|^2/\Psi(t,x)\,dV(x,S)$$

$$- \int D_x^2\Psi(t,x)\cdot S\,dV(x,S).$$

Since Ψ is a barrier function, we have for appropriate $\phi\colon \underline{\underline{R}} \to \underline{\underline{R}}^+$ (assuming $a = 0$ without loss of generality)

$$(3) \qquad \Psi(t,x) = \phi(|x|^2 + 2kt),$$

$$(4) \qquad \partial\Psi(t,x)/\partial t = 2k\phi'(|x|^2 + 2kt),$$

$$(5) \qquad D_x\Psi(t,x) = 2\phi'(|x|^2 + 2kt)x, \quad \text{and}$$

$$(6) \qquad D_x^2\Psi(t,x) = 4\phi''(|x|^2 + 2kt)x \otimes x + 2\phi'(|x|^2 + 2kt)\underline{\underline{I}}.$$

Hence, dropping the negative square from (2) and using (3), (5), and (6), we get

$$\delta(V,\Psi(t,\cdot))(\underline{\underline{h}}(V,\cdot))$$

$$\leq \int_{\{(x,S):\Psi(t,x)>0\}} |S(x)|^2|\phi'(|x|^2 + 2kt)|^2/\phi(|x|^2 + 2kt)$$

$$- 4|S(x)|^2\phi''(|x|^2 + 2kt) - 2k\phi'(|x|^2 - 2kt)\,dV(x,S).$$

Since a barrier function is defined so that $|\phi'|^2/\phi \leq 4\phi''$, we have by (4)

$$\delta(V,\Psi(t,\cdot))(\underline{\underline{h}}(V,\cdot)) \leq \int -2k\phi'(|x|^2 + 2kt)\,dV(x,S)$$

$$\leq -\|V\|(\partial\Psi(t,\cdot)/\partial t).$$

The theorem now follows from 3.5. □

3.7. Sphere barrier to external varifolds.

Theorem: If V_t is a varifold moving by its mean curvature, $R > 0$, and $\|V_0\|\underline{\underline{U}}(0,R) = 0$, then

$$\|V_t\|\underline{\underline{U}}(0,(R^2 - 2kt)^{1/2}) = 0$$

for $0 \leq t \leq R^2/2k$.

Proof: Define $\phi: \underline{\underline{R}} \to \underline{\underline{R}}^+$ by

$$\phi(r) = \begin{cases} (R^2 - r)^4 & \text{for } r \leq R^2, \\ 0 & \text{for } r > R^2. \end{cases}$$

Since $\phi'(r)^2 \leq 4\phi(r)\phi''(r)$ for all $r \in \underline{\underline{R}}$, we can define a barrier function $\Psi(t,x) = \phi(|x|^2 + 2kt)$. By 3.6,

$$D^+\|V_t\|(\Psi(t,\cdot)) \leq 0$$

for each $t \in \underline{R}^{+}$. Since $\|V_0\| (\Psi(0,\cdot)) = 0$ by hypothesis, we have $\|V_t\| (\Psi(t,\cdot)) = 0$ and thus

$$\|V_t\| \underline{U}(0, (R^2 - 2kt)^{1/2}) = 0$$

for $0 \leq t \leq R^2/2k$. $\qquad\qquad\qquad\qquad\qquad\qquad\square$

Remark: Obviously, by time and space translation invariance, the theorem remains true for initial times other than $t = 0$ and centers other than the origin.

3.8. Convex set barriers.

Theorem: If V_t is a varifold moving by its mean curvature, K is a closed convex subset of R^n, and $\mathrm{spt}\|V_0\| \subset K$ then $\mathrm{spt}\|V_t\| \subset K$ for all $t > 0$.

Proof: Suppose $\|V_t\| (\underline{R}^n \sim K) > 0$ for some $t > 0$. Then one could find a ball $\underline{U}(a,r) \subset \underline{R}^n \sim K$ such that $\|V_t\| \underline{U}(a,r) > 0$ and

$$\underline{U}(a, (r^2 + 2kt)^{1/2}) \subset \underline{R}^n \sim K.$$

But by hypothesis

$$\|V_0\| \underline{U}(a, (r^2 + 2kt)^{1/2}) = 0,$$

so by 3.7 we have $\|V_t\| \underline{U}(a,r) = 0$, which is a contradiction to $\|V_t\| \underline{U}(a,r) > 0$. $\qquad\qquad\qquad\qquad\qquad\square$

3.9. Sphere barrier to internal varifolds.

Theorem: *If* V_t *is a varifold moving by its mean curvature,* $R > 0$, *and* $\operatorname{spt}\|V_0\| \subset \underline{B}(0,R)$, *then*

$$\operatorname{spt}\|V_t\| \subset \underline{B}(0,(R^2 - 2kt)^{1/2})$$

for $0 \le t \le R^2/2k$ *and* $V_t = 0$ *for* $t > R^2/2k$.

Proof: Let Ψ be the barrier function generated by

$$\phi(r) = \begin{cases} 0 & r < R^2, \\ (r - R^2)^4 & r \ge R^2. \end{cases}$$

By 3.8, $\operatorname{spt}\|V_t\| \subset \underline{B}(0,R)$ for all $t \ge 0$, so the support of Ψ can be made compact for $t \le R^2/k$ without affecting its properties with respect to V_t. By hypothesis we have $\|V_0\|(\Psi(0,\cdot)) = 0$, and by 3.6 we have

$$D^+\|V_t\|(\Psi(t,\cdot)) \le 0$$

for $t \le R^2/k$. Hence $\|V_t\|(\Psi(t,\cdot)) = 0$ for $t \le R^2/2k$, and the conclusion follows because $\Psi(t,x) > 0$ for all x for all $t > R^2/2k$. □

3.10. Continuity properties of $\|V_t\|$.

Theorem: *Suppose* V_t *is a varifold moving by its mean curvature and* $\psi \in \underline{C}_0^2(\underline{R}^n, \underline{R}^+)$. *Then*

(a) $$\lim_{s \uparrow t} \|V_s\|(\psi) \geq \|V_t\|(\psi) \geq \lim_{s \downarrow t} \|V_s\|(\psi)$$

for all $t \in \underline{R}^+$,

(b) $\|V_t\|(\psi)$ *is a continuous and differentiable function of* t *at almost all* $t \in \underline{R}^+$,

(c) $\|V_t\|$ *is a continuous function of* t *at almost all* $t \in \underline{R}^+$.

Proof: Suppose $T > 0$. By 2.1, $|D\psi|^2/\psi$ is bounded with compact support, and therefore we may construct a barrier function $\Psi: [0,T] \times \underline{R}^n \to \underline{R}^+$ such that Ψ has compact support and

$$|D\psi|^2/\psi \leq \Psi(t,\cdot)$$

for each $t \in [0,T]$. By 3.3(1), 3.4, and 3.6 we have

$$\begin{aligned}
\overline{D}\|V_t\|(\psi) &\leq \delta(V_t,\psi)(\underline{h}(V_t,\cdot)) \\
&\leq \|V_t\|(|D\psi|^2/\psi) \\
&\leq \|V_t\|(\Psi(t,\cdot)) \\
&\leq \|V_0\|(\Psi(0,\cdot)) < \infty.
\end{aligned}$$

Conclusions (a) and (b) follow from the uniform boundedness of the upper derivate of $\|V_t\|(\psi)$ in $[0,T]$ and the arbitrariness of T.

Conclusion (c) follows since the space of test functions $\underline{\underline{C}}_0(\underline{\underline{R}}^n, \underline{\underline{R}}^+)$ has a countable dense subset from $\underline{\underline{C}}_0^2(\underline{\underline{R}}^n, \underline{\underline{R}}^+)$. □

4. Existence of varifolds moving by their mean curvature.

In this chapter we construct for a certain type of initial varifold V_0 a one parameter family of varifolds V_t defined for all $t \in \mathbb{R}^+$ and satisfying the necessary condition for motion by mean curvature given in 3.3:

$$\overline{D}\|V_t\|(\psi) \leq \delta(V_t,\psi)(\underline{h}(V_t,\cdot))$$

for any $\psi \in \mathbb{C}_0^1(\mathbb{R}^n,\mathbb{R}^+)$ and for all $t \in \mathbb{R}^+$. As 4.15 shows, in case V_0 is a smooth manifold, the construction given here agrees with the more straightforward mapping approach described in 3.1, as long as the latter works. The key properties of the present construction are proven in the last section of this chapter.

4.1. Definitions.

We wish to include noncompact surfaces in our treatment. Therefore, to keep integrals finite, we arbitrarily choose a weighting function $\Omega \in \mathbb{C}^3(\mathbb{R}^n,\mathbb{R}^+)$ satisfying the conditions $|D\Omega(x)| < \Omega(x)$ and $\|D^2\Omega(x)\| < \Omega(x)$ for all $x \in \mathbb{R}^n$. Note that Ω is never zero. Define the Ω-norm on $\mathbb{C}(\mathbb{G}_k(\mathbb{R}^n),\mathbb{R})$ by

$$\|\phi\|_\Omega = \sup\{|\phi(x,S)|/\Omega(x) : (x,S) \in \mathbb{G}_k(\mathbb{R}^n)\}$$

and define the normed linear space

$$\Omega\mathbb{C}(\mathbb{G}_k(\mathbb{R}^n)) = \{\phi \in \mathbb{C}(\mathbb{G}_k(\mathbb{R}^n),\mathbb{R}) : \|\phi\|_\Omega < \infty\}.$$

Then the set of positive continuous linear functionals on $\Omega C(G_k(R^n))$ is

$$\Omega V = \{V \in V_k(R^n): \|V\|(\Omega) < \infty\}.$$

We shall use

$$\Omega \lim_{m\to\infty} V_m = V$$

to denote convergence in the Ω topology. Note that $\Omega \lim_{m\to\infty} V_m = V$ implies $\lim_{m\to\infty} V_m = V$ in the varifold topology defined in 2.3. Since $\Omega C(G_k(R^n))$ is separable, if $M < \infty$ then

$$\{V \in \Omega V: \|V\|(\Omega) \leq M\}$$

is compact.

We define

$$\Omega R = \Omega V \cap RV_k(R^n),$$

$$\Omega I = \Omega V \cap IV_k(R^n),$$

and we define the set of <u>initial</u> <u>varifolds</u> Ω to consist of all $V \in \Omega R$ such that

(1) $\Theta^k(\|V\|,x) \geq 1$ for $\|V\|$ almost all $x \in R^n$ and

(2) $\operatorname{spt}\|V\|$ is $\mathcal{H}^k$ locally finite.

Condition (2) is not an unreasonable restriction. Indeed, the second half of the proof of 6.13 shows that if V_t is a varifold moving by its mean curvature that satisfies (1), then the instantaneous mass loss at t is proportional to the $\mathcal{H}^k$ measure of the set of points $x \in \operatorname{spt}\|V_t\|$ with $\Theta^k(\|V_t\|,x) = 0$. Thus V_t would satisfy (2) for all $t > 0$. We require (2) to hold for V_0 because this hypothesis makes 4.16 much simpler.

It follows from (1) and (2) that $V \in \underline{\underline{\Omega}}$ is of the form

$$V = \underline{\underline{v}}(S) \llcorner \beta$$

where S is a closed countably $(\mathcal{H}^k,k)$ rectifiable subset of $\underline{\underline{R}}^n$ and $\beta\colon \underline{\underline{G}}_k(\underline{\underline{R}}^n) \to \underline{\underline{R}}^+$ is a locally $\underline{\underline{v}}(S)$ summable function with values greater than or equal to 1 $\underline{\underline{v}}(S)$ almost everywhere. *If V is also integral, then β has integral values.*

It follows from [AF I.1(13)] and the properties of Ω that if $f\colon \underline{\underline{R}}^n \to \underline{\underline{R}}^n$ is a Lipschitz map with $|f(x) - x|$ bounded, then the induced mapping $f_\#$ preserves $\underline{\underline{\Omega R}}$, $\underline{\underline{\Omega I}}$, and $\underline{\underline{\Omega}}$.

If $g \in \underline{\underline{C}}^1(\underline{\underline{R}}^n,\underline{\underline{R}}^n)$ and $\sup\{\|Dg(x)\|/\Omega(x) : x \in \underline{\underline{R}}^n\} < \infty$, then we may still define for $V \in \underline{\underline{\Omega V}}$

$$\begin{aligned}\delta V(g) &= \int Dg(x)\cdot S\,dV(x,S)\\ &= \int \underline{\underline{h}}(V,x)\cdot g(x)\,d\|V\|x \quad (\text{when } \|\delta V\|_{\text{sing}} = 0)\end{aligned}$$

and have $\Omega \lim_{m\to\infty} V_m = V$ imply $\lim_{m\to\infty} \delta V_m(g) = \delta V(g)$. If $\phi \in \underline{\underline{C}}^1(\underline{\underline{R}}^n,\underline{\underline{R}})$ and $\sup\{|\phi(x)|/\Omega(x) : x \in \underline{\underline{R}}^n\} < \infty$, then we may define $\delta(V,\phi)(\underline{\underline{h}}(V,\cdot))$ as in 3.2.

The choice of the weighting function Ω enters into the actual construction in 4.9, so for a given initial V_0, the later V_t may depend on Ω. However, there are many other places arbitrary choices are made in this construction, and the solution V_t may not be unique, as noted in 3.3. Since we are concerned with existence here, nonuniqueness does not bother us.

Some sets of test functions used in this chapter will be: for each $i \in \underline{\underline{N}}$,

(3) $$\mathscr{A}_i = \{\phi \in \underline{\underline{C}}^3(\underline{\underline{R}}^n,\underline{\underline{R}}^+) : \phi(x) \leq \Omega(x),\ |D\phi(x)| \leq i\phi(x),$$
$$\text{and } \|D^2\phi(x)\| \leq i\phi(x) \text{ for all } x \in \underline{\underline{R}}^n\}.$$

Some sets of test vectorfields will be: for each $i \in \underline{\underline{N}}$,

(4) $$\mathscr{D}_i = \{g \in \underline{\underline{C}}^2(\underline{\underline{R}}^n,\underline{\underline{R}}^n) : |g(x)| \leq i\Omega(x),\ \|Dg(x)\| \leq i\Omega(x),$$
$$\text{and } \|D^2g(x)\| \leq i\Omega(x) \text{ for all } x \in \underline{\underline{R}}^n\}.$$

4.2. Estimates on growth of test functions.

Proposition: If $i \in \underline{\underline{N}}$, $\phi \in \mathscr{A}_i$, and $g \in \mathscr{D}_i$, then for all $x, y \in \underline{\underline{R}}^n$

(i) $\phi(y) \le \phi(x) \exp i|x - y|$,

(ii) $|\phi(y) - \phi(x) - D\phi(x)\cdot(y - x)|$

$$\le [i^{-1}(\exp(i|y - x|) - 1) - |y - x|]\phi(x),$$

(iii) $|D\phi(y) - D\phi(x)| \le [(\exp i|y - x|) - 1]\phi(x)$,

(iv) $|g(y) - g(x)| \le i[(\exp|y - x|) - 1]\Omega(x)$, and

(v) $\|Dg(y) - Dg(x)\| \le i[(\exp|y - x|) - 1]\Omega(x)$.

Proof: These properties are consequences of the definitions of $\mathcal{A}_i$ and ω_i. □

4.3. The smoothed mean curvature.

In approximating motion by mean curvature, we shall need smooth approximations of the mean curvature defined for any initial varifold.

For each $0 < \varepsilon < 1/2$ define $\Phi_\varepsilon\colon \mathbb{R}^n \to \mathbb{R}^+$ by

(1) $$\Phi_\varepsilon(x) = \beta(\varepsilon)\varepsilon^{-n}\exp[-x^2/(\varepsilon^2 + \varepsilon^4|x|)],$$

where $\beta(\varepsilon)$ is defined so that

(2) $$\int \Phi_\varepsilon(x)\,dx = 1.$$

Note that for $x, y \in \mathbb{R}^n$ we have

(3) $$\Phi_\varepsilon(x - y) \le \beta(\varepsilon)\varepsilon^{-n}\exp(-|x|),$$

(4) $$|D\Phi_\varepsilon(x)| \le \varepsilon^{-4}\Phi_\varepsilon(x), \quad \text{and}$$

(5) $$\|D^2\Phi_\varepsilon(x)\| \le \varepsilon^{-8}\Phi_\varepsilon(x).$$

Hence for $V \in \underline{\underline{\Omega}}$ we may define convolutions $\Phi_\varepsilon * V$, $\Phi_\varepsilon * \|V\|$, and $\Phi_\varepsilon * \delta V$. The last two of these can also be viewed as smooth functions on $\underline{\underline{R}}^n$ defined by

(6) $$\Phi_\varepsilon * \|V\|(x) = \int \Phi_\varepsilon(y - x)\,d\|V\|y \quad \text{and}$$

(7) $$\Phi_\varepsilon * \delta V(x) = \int S(D\Phi_\varepsilon(x - y))\,dV(y,S).$$

Eq. (7) is true because for $g \in \underline{\underline{C}}_0^1(\underline{\underline{R}}^n, \underline{\underline{R}}^n)$,

$$\begin{aligned}
\int \Phi_\varepsilon * \delta V(x)\cdot g(x)\,dx &= \delta V(\Phi_\varepsilon * g) \\
&= \int D(\Phi_\varepsilon * g)\cdot S \;dV(x,S) \\
&= \int (D\Phi_\varepsilon) * g\cdot S \;dV(x,S) \\
&= \iint D\Phi_\varepsilon(y - x) \otimes g(y)\cdot S \;dy\;dV(x,S) \\
&= \int g(y) \cdot \int S(D\Phi_\varepsilon(y - x))\,dV(x,S)\,dy.
\end{aligned}$$

Clearly $\Phi_\varepsilon * \|V\| = \|\Phi_\varepsilon * V\|$, and $\Phi_\varepsilon * \delta V = \delta(\Phi_\varepsilon * V)$ because

$$(\Phi_\varepsilon * \delta V)(g) = \delta V(\Phi_\varepsilon * g)$$

$$= \int D(\Phi_\varepsilon * g)\cdot S \; dV(x,S)$$

$$= \int (\Phi_\varepsilon * Dg)\cdot S \; dV(x,S)$$

$$= \delta(\Phi_\varepsilon * V)(g).$$

From (5), (6) and (7) we get

(8) $$|\Phi_\varepsilon * \delta V(x)|/\Phi_\varepsilon * \|V\|(x)$$

$$\leq \int |D\Phi_\varepsilon(x - y)|d\|V\|y \Big/ \int \Phi_\varepsilon(x - y)d\|V\|y$$

$$\leq \varepsilon^{-4}.$$

Thus we may define the <u>smoothed</u> <u>mean</u> <u>curvature</u> of V, denoted $h_\varepsilon(V)$, to be the vectorfield

(9) $$h_\varepsilon(V) = -\Phi_\varepsilon * (\Phi_\varepsilon * \delta V/\Phi_\varepsilon * \|V\|).$$

When V is unambiguous, we shall write h_ε for $h_\varepsilon(V)$ and $h_\varepsilon(x)$ for $h_\varepsilon(V)(x)$. Proposition 4.8 shows that h_ε is in fact an approximation to the mean curvature.

It can be shown in the standard way that $\Omega \lim_{m\to\infty} V_m = V$ and $\lim_{m\to\infty} \varepsilon_m = 0$ imply

(10) $$\Omega \lim_{m\to\infty} \Phi_{\varepsilon_m} * V_m = V.$$

4.4. The smoothness of h_ε.

Proposition: If $V \in \underline{\Omega}$ and $0 < \varepsilon < 1/2$, then for all $x \in \underline{R}^n$ we have

(i) $|h_\varepsilon(x)| \le \varepsilon^{-4}$,

(ii) $\|Dh_\varepsilon(x)\| \le \varepsilon^{-8}$, and

(iii) $\|D^2 h_\varepsilon(x)\| \le \varepsilon^{-12}$.

Proof: From 4.3(9), (8), (2), (4), and (5) we get

$$|h_\varepsilon| \le \Phi_\varepsilon * |\Phi_\varepsilon * \delta V / \Phi_\varepsilon * \|V\||$$

$$\le \Phi_\varepsilon * \varepsilon^{-4} \le \varepsilon^{-4},$$

$$\|Dh_\varepsilon\| \le |D\Phi_\varepsilon| * |\Phi_\varepsilon * \delta V / \Phi_\varepsilon * \|V\|| < \varepsilon^{-8}, \text{ and}$$

$$\|D^2 h_\varepsilon\| \le \|D^2\Phi_\varepsilon\| * |\Phi_\varepsilon * \delta V / \Phi_\varepsilon * \|V\|| \le \varepsilon^{-12}. \qquad \square$$

Because of these estimates, we may use 2.10(1,2,3) to define $\delta(V,\varepsilon)(h_\varepsilon(V))$ for $\phi \in \mathscr{A}_i$.

4.5. Some constants.

Define $0 < c_1 < 1/10$ such that if $i \in \underline{N}$, $\varepsilon < c_1 i^{-1}$, and $\phi \in \mathscr{A}_i$ or $\phi \equiv 1$, then for all $j \le i$ we have

(1) $$\Phi_\varepsilon * \phi \le 2\phi,$$

(2) $$(\varepsilon^{-1}(e^{j|z|} - 1)\Phi_\varepsilon(z))*\phi \le nj\phi,$$

(3) $$(\varepsilon^{-2}(e^{j|z|} - 1)^2\Phi_\varepsilon(z))*\phi \le nj^2\phi,$$

(4) $$(\varepsilon^{-1}(e^{j|z|} - 1)^2|D\Phi_\varepsilon(z)|)*\phi \le n^2j^2\phi,$$

(5) $$(\varepsilon^{-2}j^{-2}[e^{j|z|} - 1 - j|z|]^2|D\Phi_\varepsilon(z)|^2/\Phi_\varepsilon(z))*\phi \le n^2j^2\phi, \quad \text{and}$$

(6) $$([|z| + 3\varepsilon^2|z|^2 + (2\varepsilon^4 + \varepsilon^{-2})|z|^3 + 4|z|^4]^2\Phi_\varepsilon(z))*\phi \le \phi.$$

Also define

(7) $$c_2(i,\varepsilon) = \sup\{G_\varepsilon*\phi(x)/\phi(x): x \in \underline{\underline{R}}^n, \quad \phi \in \mathscr{A}_i\}$$

where

(8) $$G_\varepsilon(x) = \begin{cases} 0 & \text{for} \quad 0 \le |x| \le \varepsilon^{1/2}, \\ |D\Phi_\varepsilon(x)| & \text{for} \quad |x| > \varepsilon^{1/2}. \end{cases}$$

We have

$$G_\varepsilon*\phi(x)/\phi(x) = \int_{|x-y|>\varepsilon^{1/2}} \phi(x)^{-1}|D\Phi_\varepsilon(y - x)|\phi(y)\,dy.$$

Since, from 4.3(1),

$$|D\Phi_\varepsilon(z)| \le 2\varepsilon^{-2}|z|\Phi_\varepsilon(z)$$

and, by 4.2(i),

$$\phi(y) \le \phi(x) \exp i|y - x|,$$

we have, using 4.3(1),

$$c_2(i,\varepsilon) \le \int_{|z|>\varepsilon^{1/2}} 2\varepsilon^{-2}|z|\beta(\varepsilon)\varepsilon^{-n}\exp[-|z|^2/(\varepsilon^2 + \varepsilon^4|z|) + i|z|]dz.$$

Thus for any $p \in \underline{\underline{R}}$, in particular for $p < 0$,

(9) $$\lim_{\varepsilon\to 0} \varepsilon^p c_2(i,\varepsilon) = 0.$$

<u>Lemma</u>: <u>If</u> $V \in \underline{\underline{\Omega}}$, $i \in \underline{\underline{N}}$, $\phi \in \mathscr{A}_i$ <u>and</u> $0 < \varepsilon < c_1 i^{-1}$ <u>then</u>

(10) $$|\langle\phi, \Phi_\varepsilon * \|V\|\rangle - \langle\phi, \|V\|\rangle| \le \varepsilon n i \|V\|(\phi).$$

<u>Proof</u>:

$$|\langle\phi, \Phi_\varepsilon * \|V\|\rangle - \langle\phi, \|V\|\rangle|$$

$$= \left|\iint \phi(y)\Phi_\varepsilon(y - x)dy - \phi(x)d\|V\|x\right|$$

$$= \left|\iint (\phi(y) - \phi(x))\Phi_\varepsilon(y - x)dy\ d\|V\|x\right|$$

$$\le \iint \{\exp[i|y - x|] - 1\}\phi(x)\Phi_\varepsilon(y - x)dy\ d\|V\|x$$

(by 4.3 (i))

$$\leq \iint \{\exp[i|y-x|] - 1\}\Phi_\varepsilon(y-x)\,dy\ \phi(x)\,d\|V\|x$$

$$\leq \varepsilon n i \|V\|(\phi) \qquad \text{(by 4.5(2))}. \qquad \square$$

4.6. Some estimates on h_ε.

Proposition: If $V \in \underline{\Omega}$, $i \in \underline{N}$, and $0 < \varepsilon < c_1 i^{-1}$ then for any $g \in \mathcal{D}_i$

(i) $$\left|\int h_\varepsilon(x)\cdot g(x)\,d\|V\|x + \int \Phi_\varepsilon * \delta V(x)\cdot g(x)\,dx\right|$$

$$\leq n i \varepsilon \langle \Omega, |\Phi_\varepsilon * \delta V|^2/\Phi_\varepsilon * \|V\|\rangle^{1/2} \|V\|(\Omega)^{1/2},$$

(ii) $$\left|\int S^\perp(g(x))\cdot h_\varepsilon(x)\,dV(x,S)\right.$$

$$\left. + \iint S^\perp(g(x))\,d(\Phi_\varepsilon * V)^{(x)}S\cdot\Phi_\varepsilon * \delta V(x)\,dx\right|$$

$$\leq n i \varepsilon \langle \Omega, |\Phi_\varepsilon * \delta V|^2/\Phi_\varepsilon * \|V\|\rangle^{1/2} \|V\|(\Omega)^{1/2},$$ and

(iii) if $\phi \in \mathcal{A}_i$ and $g = D\phi$ then one may replace Ω by ϕ in the right hand sides of (i) and (ii).

Proof: It follows from the definition of h_ε in 4.3(9) that

$$\int h_\varepsilon(x)\cdot g(x)\,d\|V\|x = \langle g\|V\|, -\Phi_\varepsilon * (\Phi_\varepsilon * \delta V/\Phi_\varepsilon * \|V\|)\rangle$$

$$= -\langle \Phi_\varepsilon * (g\|V\|), \Phi_\varepsilon * \delta V/\Phi_\varepsilon * \|V\|\rangle,$$

and we can write

$$\int \Phi_\varepsilon * \delta V(x) \cdot g(x)\,dx = \langle g\Phi_\varepsilon * \|V\|,\ \Phi_\varepsilon * \delta V / \Phi_\varepsilon * \|V\| \rangle.$$

Therefore

(1) $$\left| \int h_\varepsilon(x) \cdot g(x)\,d\|V\|x + \int \Phi_\varepsilon * \delta V(x) \cdot g(x)\,dx \right|$$

$$\leq \langle |\Phi_\varepsilon * (g\|V\|) - g\Phi_\varepsilon * \|V\||,\ |\Phi_\varepsilon * \delta V| / \Phi_\varepsilon * \|V\| \rangle.$$

Now for each $x \in \underline{\underline{R}}^n$, using 4.2 (iv) and Schwarz' inequality,

(2) $$|\Phi_\varepsilon * (g\|V\|)(x) - g(x)\Phi_\varepsilon * \|V\|(x)|^2$$

$$= \left| \int (g(y) - g(x))\Phi_\varepsilon(y - x)\,d\|V\|y \right|^2$$

$$\leq \left| \int i[(\exp|y - x|) - 1]\Omega(x)\Phi_\varepsilon(y - x)\,d\|V\|y \right|^2$$

$$\leq i^2\Omega(x)^2 \int [(\exp|x - y|) - 1]^2\Phi_\varepsilon(y - x)\,d\|V\|y \int \Phi_\varepsilon(y - x)\,d\|V\|y$$

$$= i^2\Omega(x)^2([\exp|z| - 1]^2\Phi_\varepsilon(z)) * \|V\|(x)\,\Phi_\varepsilon * \|V\|(x).$$

Using Schwarz' inequality on (1) and then using (2) gives

(3) $$\left| \int h_\varepsilon(x) \cdot g(x)\,d\|V\|x + \int \Phi_\varepsilon * \delta V(x) \cdot g(x)\,dx \right|^2$$

$$\leq \langle \Omega, |\Phi_\varepsilon * \delta V|^2 / \Phi_\varepsilon * \|V\| \rangle \langle \Omega^{-1}, |\Phi_\varepsilon * (g\|V\|) - g\Phi_\varepsilon * \|V\||^2 / \Phi_\varepsilon * \|V\| \rangle$$

$$\leq \langle \Omega, |\Phi_\varepsilon * \delta V|^2 / \Phi_\varepsilon * \|V\| \rangle \langle \Omega^{-1}, i^2\Omega^2([\exp|z| - 1]^2\Phi_\varepsilon(z)) * \|V\| \rangle$$

$$\leq \langle \Omega, |\Phi_\varepsilon * \delta V|^2 / \Phi_\varepsilon * \|V\| \rangle i^2 \langle ([\exp|z| - 1]^2\Phi_\varepsilon(z)) * \Omega, \|V\| \rangle$$

$$\leq ni^2\varepsilon^{-2}\langle\Omega, |\Phi_\varepsilon * \delta V|^2/\Phi_\varepsilon * \|V\|\rangle\|V\|(\Omega),$$

where at the end we used 4.5(3), with $j = 1$. This proves (i).

For (ii) we have

$$(4) \qquad \int S^\perp(g(x))\cdot h_\varepsilon(x)\,dV(x,S)$$

$$= -\iint S^\perp(g(x))\cdot \Phi_\varepsilon(y - x)\,\Phi_\varepsilon * \delta V(y)/\Phi_\varepsilon * \|V\|(y)\,dy\,dV(x,S)$$

and

$$(5) \qquad \iint S^\perp(g(x))\,d(\Phi_\varepsilon * V)^{(x)}S\cdot \Phi_\varepsilon * \delta V(x)\,dx$$

$$= \int S^\perp(g(x))\cdot \Phi_\varepsilon * \delta V(x)/\Phi_\varepsilon * \|V\|(x)\,d\Phi_\varepsilon * V(x,S)$$

$$= \iint S^\perp(g(y))\cdot \Phi_\varepsilon(y - x)\,\Phi_\varepsilon * \delta V(y)/\Phi_\varepsilon * \|V\|(y)\,dy\,dV(x,S).$$

Adding (4) and (5) gives

$$\left|\int S^\perp(g(x))\cdot h_\varepsilon(x)\,dV(x,S) + \iint S^\perp(g(x))\,d(\Phi_\varepsilon * V)^{(x)}S\cdot \Phi_\varepsilon * \delta V(x)\,dx\right|$$

$$\leq \iint S^\perp(g(y) - g(x))\cdot \Phi_\varepsilon(y - x)\,\Phi_\varepsilon * \delta V(y)/\Phi_\varepsilon * \|V\|(y)\,dy\,dV(x,S)$$

$$\leq \iint |g(y) - g(x)|\,\Phi_\varepsilon(y - x)\,d\|V\|x\ \ \Phi_\varepsilon * \delta V(y)/\Phi_\varepsilon * \|V\|(y)\,dy$$

which can be treated exactly the same way as (1) was to give the same estimate as in (i).

If $\phi \in \mathscr{A}_i$ and $g = D\phi$, then in the above derivations we can substitute ϕ for Ω in the first inequality of (3), use

4.2(iii) instead of 4.2(iv) in (2), and we use 4.5(3) with $j = i$. The net result is just to replace Ω with ϕ in the right hand sides of (i) and (ii). □

4.7. More estimates on h_ε.

Proposition: If $V \in \underline{\Omega}$, $i \in \underline{N}$, $\phi \in \mathcal{A}_i$, and $0 < \varepsilon < c_1 i^{-1}$, then

(i) $$|\delta V(\phi h_\varepsilon) + \langle \phi, |\Phi_\varepsilon * \delta V|^2 / \Phi_\varepsilon * \|V\| \rangle|$$

$$\leq 3ni\varepsilon^{1/2} \langle \phi, |\Phi_\varepsilon * \delta V|^2 / \Phi_\varepsilon * \|V\| \rangle + ni[2\varepsilon + \varepsilon^{-6} c_2(i,\varepsilon)] \|V\|(\phi),$$

(ii) $$\int |h_\varepsilon(x)|^2 \phi(x) \, d\|V\| x \leq (1 + \varepsilon ni) \langle \phi, |\Phi_\varepsilon * \delta V|^2 / \Phi_\varepsilon * \|V\| \rangle.$$

Proof: We have from 2.6

$$\begin{aligned}
\delta V(\phi h_\varepsilon) &= \int D[\phi h_\varepsilon](x) \cdot S \, dV(x,S) \\
&= \int \phi(x) Dh_\varepsilon(x) \cdot S \, dV(x,S) + \int D\phi(x) \otimes h_\varepsilon(x) \cdot S dV(x,S) \\
&= - \iint \phi(x) D\Phi_\varepsilon(y - x) \otimes \Phi_\varepsilon * \delta V(y) / \Phi_\varepsilon * \|V\|(y) \cdot S dy dV(x,S) \\
&\quad + \iint \Phi_\varepsilon(y - x) \Phi_\varepsilon * \delta V(y) / \Phi_\varepsilon * \|V\|(y) \otimes D\phi(x) \cdot S dy dV(x,S),
\end{aligned}$$

and using 4.3(7),

$$\langle \phi, |\Phi_\varepsilon * \delta V|^2 / \Phi_\varepsilon * \|V\| \rangle$$

$$= \int \phi(y)\Phi_\varepsilon * \delta V(y) \cdot \Phi_\varepsilon * \delta V(y)/\Phi_\varepsilon * \|V\|(y)\,dy$$

$$= \iint \phi(y) S(D\Phi_\varepsilon(y - x)) \cdot \Phi_\varepsilon * \delta V(y)/\Phi_\varepsilon * \|V\|(y)\,dV(x,S)\,dy.$$

Thus

$$(1) \qquad |\delta V(\phi h_\varepsilon) + \langle \phi, |\Phi_\varepsilon * \delta V|^2/\Phi_\varepsilon * \|V\| \rangle|$$

$$\leq \left| \iint -\phi(x) S(D\Phi_\varepsilon(y - x)) + S(D\phi(x))\Phi_\varepsilon(y - x) \right.$$

$$\left. + \phi(y) S(D\Phi_\varepsilon(y - x))\,dV(x,S) \cdot \Phi_\varepsilon * \delta V(y)/\Phi_\varepsilon * \|V\|(y)\,dy \right|.$$

We shall work on the inner integral, first approximating $\phi(y) - \phi(x)$ by $D\phi(y) \cdot (y - x)$. The maximum error we are making is

$$\iint |\phi(y) - \phi(x) - D\phi(y) \cdot (y - x)|\,|D\Phi_\varepsilon(y - x)|\,dV(x,S)$$

$$\cdot |\Phi_\varepsilon * \delta V(y)|/\Phi_\varepsilon * \|V\|(y)\,dy,$$

which by 4.2(ii) is less than

$$\iint [i^{-1}(\exp(i|y - x|) - 1) - |y - x|]\phi(y)\,|D\Phi_\varepsilon(y - x)\ d\|V\|x$$

$$\cdot |\Phi_\varepsilon * \delta V(y)|/\Phi_\varepsilon * \|V\|(y)\,dy$$

which, using Schwarz' inequality, is less than

$$\left\{\iint \left| \int [i^{-1}(\exp i|y-x| - 1) - |y-x|]\,|D\Phi_\varepsilon(y-x)|\,d\|V\|x\right|^2 \right.$$

$$\left. \cdot\, \phi(y)/\Phi_\varepsilon * \|V\|(y)\,dy\right\}^{1/2}$$

$$\cdot \left\{\int |\Phi_\varepsilon * \delta V(y)|^2/\Phi_\varepsilon * \|V\|(y)\,\phi(y)\,dy\right\}^{1/2}$$

which, again using Schwarz' inequality, is less than

$$\left\{\iint [i^{-1}(\exp i|y-x| - 1) - |y-x|]^2 |D\Phi_\varepsilon(y-x)|^2/\Phi_\varepsilon(y-x)\,d\|V\|x\right.$$

$$\left. \cdot \int \Phi_\varepsilon(y-x)\,d\|V\|x \;\; \phi(y)/\Phi_\varepsilon * \|V\|(y)\,dy\right\}^{1/2}$$

$$\cdot \;\; \langle \phi, |\Phi_\varepsilon * \delta V|^2/\Phi_\varepsilon * \|V\|\,\rangle^{1/2},$$

which, cancelling $\Phi_\varepsilon * \|V\|(y)$, interchanging the order of integration, and writing as a convolution, is less than

$$\langle \{[i^{-1}(\exp i|z| - 1) - |z|]^2 |D\Phi_\varepsilon(z)|^2/\Phi_\varepsilon(z)\} * \phi, \|V\|\,\rangle^{1/2}$$

$$\cdot \langle \phi, |\Phi_\varepsilon * \delta V|^2/\Phi_\varepsilon * \|V\|\,\rangle^{1/2},$$

which, using 4.5(5) with $j = i$, is less than

$$(2) \qquad n i \varepsilon \|V\|(\phi)^{1/2} \langle \phi, |\Phi_\varepsilon * \delta V|^2/\Phi_\varepsilon \|V\|\,\rangle^{1/2}.$$

Next, in (1) we approximate $D\phi(x)$ by $D\phi(y)$. The maximum error we are making is

$$\iint |D\phi(y) - D\phi(x)|\,\Phi_\varepsilon(y-x)\,d\|V\|x \cdot |\Phi_\varepsilon * \delta V(y)|/\Phi_\varepsilon * \|V\|(y)\,dy,$$

which, using 4.2(iii) is less than

$$\iint [(\exp i|y - x|) - 1]\phi(y)\Phi_\varepsilon(y - x)\,d\|V\|x$$

$$\cdot |\Phi_\varepsilon * \delta V(y)|/\Phi_\varepsilon * \|V\|(y)\,dy,$$

which, using Schwarz' inequality twice, is less than

$$\left\{\int \left| \int [(\exp i|y - x|) - 1]\Phi_\varepsilon(y - x)\,d\|V\|x \right|^2 \cdot \phi(y)/\Phi_\varepsilon * \|V\|(y)\,dy\right\}^{1/2}$$

$$\cdot \left\{\int \phi(y)|\Phi_\varepsilon * \delta V(y)|^2/\Phi_\varepsilon * \|V\|(y)\,dy\right\}^{1/2}$$

$$\leq \left\{\iiint [(\exp i|y - x|) - 1]^2\Phi_\varepsilon(y - x)\,d\|V\|x\right.$$

$$\left.\cdot \int \Phi_\varepsilon(y - x)\,d\|V\|(x)\,\phi(y)/\Phi_\varepsilon * \|V\|(y)\,dy\right\}^{1/2}$$

$$\cdot \langle \phi, |\Phi_\varepsilon * \delta V|^2/\Phi_\varepsilon * \|V\|\rangle^{1/2},$$

which, cancelling $\Phi_\varepsilon * \|V\|(y)$, reversing the order of integration, and writing as a convolution, is less than

$$\langle([\exp i|z|) - 1]^2\Phi_\varepsilon(z)) * \phi, \|V\|\rangle^{1/2}\langle \phi, |\Phi_\varepsilon * \delta V|^2/\Phi_\varepsilon * \|V\|\rangle^{1/2},$$

which by 4.5(3) with $j = i$ is less than

$$n\varepsilon i\|V\|(\phi)^{1/2}\langle \phi, |\Phi_\varepsilon * \delta V|^2/\Phi_\varepsilon * \|V\|\rangle^{1/2}. \tag{3}$$

We now have the inner integral of (1) converted to

$$\int D\phi(y)\cdot(y - x)S(D\Phi_\varepsilon(y - x)) + S(D\phi(y))\Phi_\varepsilon(y - x)\,dV(x,S),$$

which is equal to, noting that $D\phi(y)$ is constant with respect to x,

$$(4) \qquad - \int S(D_x[D\phi(y)\cdot(y - x)\Phi_\varepsilon(y - x)])\,dV(x,S).$$

Next, we approximate $(y - x)\Phi_\varepsilon(y - x)$ in (4) by $-(\varepsilon^2/2)D\Phi_\varepsilon(y - x)$. It can be shown from the formula 4.3(1) for Φ_ε that

$$x\Phi_\varepsilon(x) + (\varepsilon^2/2)D\Phi_\varepsilon(x)$$

$$= (\varepsilon^2/2)x|x|(2\varepsilon^2|x| - 1)(1 + \varepsilon^2|x|)^{-2}\Phi_\varepsilon(x),$$

and one can calculate

$$|D_x\Big[D\phi(y)\cdot(y - x)\Phi_\varepsilon(y - x)\Big] + \Big[D_x\ D\phi(y)\cdot\tfrac{1}{2}\ \varepsilon^2 D\Phi_\varepsilon(y - x)\Big]|$$

$$\leq (\varepsilon^2/2)\,|D\phi(y)|\Big[2|y-x| + 6\varepsilon^2|y-x|^2 + (4\varepsilon^4 + 2\varepsilon^{-2})|y-x|^3 + 4|y-x|^4\Big]\Phi_\varepsilon$$

$$\leq i\varepsilon^2\phi(y)\Big[|y-x| + 3\varepsilon^2|y-x|^2 + (2\varepsilon^4 + \varepsilon^{-2})|y-x|^3 + 4|y-x|^4\Big]\Phi_\varepsilon(y-x).$$

The maximum error in making this approximation comes out, using 4.5(6), to be less than

$$(5) \qquad ni\varepsilon\|V\|(\phi)^{1/2}\langle\phi, |\Phi_\varepsilon * \delta V|^2/\Phi_\varepsilon * \|V\|\rangle^{1/2}.$$

By reasoning as in 4.3(7), we may write (4) as

$$[D\phi(y)\cdot z\Phi_\varepsilon(z)] * \delta V(y),$$

which after the preceding approximation becomes

$$(\varepsilon^2/2)\,[D\phi(y)\cdot D\Phi_\varepsilon(z)]*\delta V(y),$$

which it will be helpful to write in component form as

$$(\varepsilon^2/2)\sum_{j=1}^{n} D_j\phi(y)\,[D_j\Phi_\varepsilon(z)]*\delta V(y)$$

$$= -(\varepsilon^2/2)\sum_{j=1}^{n} D_j\phi(y)\,D_j[\Phi_\varepsilon*\delta V](y).$$

Then (1) becomes, after these approximations, less than

$$(\varepsilon^2/2)\left|\sum_{j=1}^{n} \langle D_j\phi D_j[\Phi_\varepsilon*\delta V],\Phi_\varepsilon*\delta V/\Phi_\varepsilon*\|V\|\rangle\right|$$

$$= (\varepsilon^2/4)\left|\sum_{j=1}^{n} \langle D_j\phi/\Phi_\varepsilon*\|V\|, D_j[\Phi_\varepsilon*\delta V]^2\rangle\right|.$$

Now we integrate by parts to get

$$(\varepsilon^2/4)\sum_{j=1}^{n} \langle D_jD_j\phi/\Phi_\varepsilon*\|V\| - D_j\ \cdot D_j\Phi_\varepsilon*\|V\|/(\Phi_\varepsilon*\|V\|)^2,(\Phi_\varepsilon*\delta V)^2\rangle|,$$

which is less than, using the properties of ϕ from 4.1,

(6) $$n\varepsilon^2\langle i\phi + i\phi|D\Phi_\varepsilon|*\|V\|/\Phi_\varepsilon*\|V\|,(\Phi_\varepsilon*\delta V)^2/\Phi_\varepsilon*\|V\|\rangle.$$

To estimate this we write $|D\Phi_\varepsilon(x)| = F_\varepsilon(x) + G_\varepsilon(x)$, where

$$F_\varepsilon(x) = \begin{cases} |D\Phi_\varepsilon(x)| & \text{for } 0 \le |x| \le \varepsilon^{1/2} \\ 0 & \text{for } |x| > \varepsilon, \end{cases}$$

$$G_\varepsilon(x) = \begin{cases} 0 & \text{for } 0 \le |x| \le \varepsilon^{1/2} \\ |D\Phi_\varepsilon(x)| & \text{for } |x| > \varepsilon. \end{cases}$$

Since $|D\Phi_\varepsilon(x)| \le 2\varepsilon^{-2}|x|\Phi_\varepsilon(x)$, we have

$$F_\varepsilon(x) \le 2\varepsilon^{-3/2}\Phi_\varepsilon(x).$$

Thus (6) is less than

$$\text{(7)} \qquad n\varepsilon^2 i\langle\phi,(\Phi_\varepsilon * \delta V)^2/\Phi_\varepsilon * \|V\|\rangle$$

$$+ n\varepsilon^2 i\langle\phi 2\varepsilon^{-3/2}\Phi_\varepsilon * \|V\|/\Phi_\varepsilon * \|V\|,(\Phi_\varepsilon * \delta V)^2/\Phi_\varepsilon * \|V\|\rangle$$

$$+ n\varepsilon^2 i\langle\phi G_\varepsilon * \|V\|,(\Phi_\varepsilon * \delta V/\Phi_\varepsilon * \|V\|)^2\rangle.$$

Since 4.3(8) says $|\Phi_\varepsilon * \delta V|/\Phi_\varepsilon * \|V\| < \varepsilon^{-4}$, we have (7) less than

$$\text{(8)} \qquad ni(\varepsilon^2 + 2\varepsilon^{1/2})\langle\phi,(\Phi_\varepsilon * \delta V)^2/\Phi_\varepsilon * \|V\|\rangle + n\varepsilon^2 i\langle G_\varepsilon * \|V\|,\phi\varepsilon^{-8}\rangle.$$

The last part of (8) can be written as

$$n\varepsilon^2 i\langle G_\varepsilon * \|V\|,\phi\varepsilon^{-8}\rangle = n\varepsilon^{-6} i\langle\|V\|,G_\varepsilon * \phi\rangle,$$

and by 4.5(7) this is less than

(9) $$n\varepsilon^{-6} i\ c_2(i,\varepsilon)\,\|V\|(\phi).$$

Adding together (2), (3), (5), (8), and (9) gives

$$|\delta V(\phi h_\varepsilon) + \langle \phi, |\Phi_\varepsilon * \delta V|^2/\Phi_\varepsilon * \|V\|\rangle|$$

$$\leq [ni\varepsilon + ni\varepsilon + ni\varepsilon]\langle \phi, |\Phi_\varepsilon * \delta V|^2/\Phi_\varepsilon * \|V\|\rangle^{1/2}\|V\|(\phi)^{1/2}$$

$$+ ni(\varepsilon^2 + 2\varepsilon^{1/2})\langle \phi, |\Phi_\varepsilon * \delta V|^2/\Phi_\varepsilon * \|V\|\rangle + ni\varepsilon^{-6} c_2(i,\varepsilon)\|V\|(\phi).$$

Applying Minkowski's inequality and recognizing that $\varepsilon^{1/2}$ dominates ε and ε^2 yields (i).

To prove (ii), note that for any convolutable function f, we have by Schwarz' inequality

$$\phi(x)|\Phi_\varepsilon * f(x)|^2 = |\int \phi(x)^{1/2}\Phi_\varepsilon(y - x)f(y)dy|^2$$

$$\leq \int \phi(y)\Phi_\varepsilon(y - x)\ f(y)|^2 dy \int \Phi_\varepsilon(y - x)\phi(x)/\phi(y)dy$$

$$\leq \Phi_\varepsilon * (\phi f^2) \int \Phi_\varepsilon(y - x)\exp[i|y - x|]dy \quad \text{(by 4.2(i))}$$

$$\leq \Phi_\varepsilon * (\phi f^2)(1 + \varepsilon n i),$$

where we have used 4.5(2) with $\phi \equiv 1$. Hence

$$\langle \phi | \Phi_\varepsilon * (\Phi_\varepsilon * \delta V / \Phi_\varepsilon * \|V\|) |^2, \|V\| \rangle$$

$$\leq \langle \Phi_\varepsilon * (\phi | \Phi_\varepsilon * \delta V / \Phi_\varepsilon * \|V\| |^2), \|V\| \rangle (1 + \varepsilon ni)$$

$$= \langle \phi | \Phi_\varepsilon * \delta V |^2 / | \Phi_\varepsilon * \|V\| |^2, \ \Phi_\varepsilon * \|V\| \rangle (1 + \varepsilon ni)$$

$$= (1 + \varepsilon ni) \langle \phi, | \Phi_\varepsilon * \delta V |^2 / \Phi_\varepsilon * \|V\| \rangle.$$

□

4.8. Showing that h_ε is an approximation of mean curvature.

Proposition: If $B < \infty$, $i \in \underline{N}$, $0 < \varepsilon < c_1 i^{-1}$, $V \in \underline{\Omega}$, $\langle \Omega, |\Phi_\varepsilon * \delta V|^2 / \Phi_\varepsilon * \|V\| \rangle < B$, $\|V\|(\Omega) < B$, and $g \in \mathscr{X}_i$, then

$$\left| \int h_\varepsilon(x) \cdot g(x) \, d\|V\| x + \delta V(g) \right| < 2ni\varepsilon B.$$

Proof: From 4.6(i) we have

$$(1) \qquad \left| \int h_\varepsilon(x) \cdot g(x) \, d\|V\| x + \int \Phi_\varepsilon * \delta V(x) \cdot g(x) \, dx \right| \leq ni\varepsilon B.$$

We also have

$$(2) \qquad \left| \int \Phi_\varepsilon * \delta V(x) \cdot g(x) \, dx - \delta V(g) \right| = |\delta V(\Phi_\varepsilon * g) - \delta V(g)|$$

$$= \left| \int S(D\Phi_\varepsilon * g(x) - Dg(x)) \, dV(x,S) \right|$$

$$\leq \int |\Phi_\varepsilon * Dg(x) - Dg(x)| \, d\|V\| x$$

$$\leq \int \left| \int \Phi_\varepsilon(y - x) Dg(y) \, dy - Dg(x) \right| d\|V\| x$$

$$\leq \int \left| \int \Phi_\varepsilon(y - x)(dg(y) - Dg(x)) \, dy \right| d\|V\| x$$

$$\leq \iint \Phi_\varepsilon(y - x) i[(\exp y - x) - 1]\Omega(x) dy \, d\|V\| x$$

$$\leq ni\varepsilon \|V\|(\Omega) < ni\varepsilon B,$$

where we used 2.6(3), 4.2(v), and 4.5(2). Combining (1) and (2) gives the desired result. □

4.9. Approximate motion by mean curvature.

The basic idea is to let a varifold move in tiny steps along the smoothed mean curvature, recalculating the smooth mean curvature at each step. Then we let ε go to zero and take a limit of approximations. However, this straightforward approach is inadequate to get the limit to be a solution of the original problem. Therefore I introduce a second type of step that takes care of all the loose ends. This section describes the two types of steps.

For each $m \in \mathbb{N}$ pick $\varepsilon(m) > 0$ such that

(1) $$\varepsilon(m) < c_1 m^{-20} n^{-2}, \quad m\varepsilon(m)k! < 1, \quad \text{and}$$

$$ni\varepsilon(m)^{-6} c_2(i,\varepsilon) < m^{-4} \quad \text{for} \quad i \leq m.$$

Define $\sigma(m) = m^{-2}$ and $\Delta t(m) = 2^{-p}$, where $p \in \mathbb{N}$ is chosen so that $\Delta t(m) < \varepsilon(m)^{60}$.

For $V \in \Omega$ and $\sigma, w > 0$, define $\mathbb{E}(V,\sigma,w)$ to be the set of all Lipschitz functions $f_1\colon \mathbb{R}^n \to \mathbb{R}^n$ such that

(a) $|f_1(x) - x| \leq \sigma$ for all $x \in \underline{\underline{R}}^n$ and

(b) $\|f_{1\#}V\|(\phi) \leq \|V\|(\phi)$ for all $\phi \in \mathscr{A}_w$.

Note that $f_1(x) = x$ satisfies (a) and (b), so $\underline{\underline{E}}(V,\sigma,w)$ is nonempty. It follows from (a) and 2.8 that $f_{1\#}$ preserves $\underline{\underline{\Omega}}$ and $\underline{\underline{I\Omega}}$. For $\phi \in \mathscr{A}_m$ we shall denote

$$\Delta_{\sigma,w}\|V\|(\phi) = \inf\{\|f_{1\#}V\|(\phi) - \|V\|(\phi): f_1 \in \underline{\underline{E}}(V,\sigma,w)\}.$$

For $V \in \underline{\underline{\Omega}}$ and $m \in \underline{\underline{N}}$ define $f_2: \underline{\underline{R}}^n \to \underline{\underline{R}}^n$ by

$$f_2(x) = x + \Delta t(m) h_{\varepsilon(m)}(v)(x).$$

From 4.4 we conclude that f_2 is a Lipschitz map with $|f_2(x) - x|$ bounded, so $f_{2\#}$ also preserves $\underline{\underline{\Omega}}$ and $\underline{\underline{I\Omega}}$.

Remark 1: The mapping f_2 approximates motion by mean curvature. The mappings f_1 are meant to do away with irregularities that are too small to be detected by the smoothed mean curvature. Condition (b) guarantees that f_1 does not do too much.

Remark 2: One can model different processes by fiddling with the first type of mapping. For example, instead of the varifold mapping defined in 2.8 one could define $f_\# V = \underline{\underline{v}}(f(\operatorname{spt}\|V\|)) \in \underline{\underline{\Omega}}$. This definition produces varifolds with density 1 everywhere and would be appropriate for modeling soap films and other instances with uniform surfaces. This model can be called the <u>reduced mass model</u>.

One could also require that f_1 be a homotopy. This would be of interest if there were dimensional obstructions to moving surfaces avoiding each other by making slight detours.

All the results of this chapter hold for all such models as long as conditions (a) and (b) are satisfied.

4.10. <u>Sufficient condition for belonging to</u> $\underline{\underline{E}}(V,\sigma,w)$.

<u>Lemma</u>: <u>If</u> $V \in \underline{\underline{\Omega}}$, $\sigma, w > 0$, B <u>is a closed subset of</u> $\underline{\underline{R}}^n$, <u>and</u> $f\colon \underline{\underline{R}}^n \to \underline{\underline{R}}^n$ <u>is a Lipschitz map such that</u>

(i) $\{x\colon f(x) \neq x\} \cup \{f(x)\colon f(x) \neq x\} \subset B$,

(ii) $|f(x) - x| < \sigma$ <u>for all</u> $x \in R^n$, <u>and</u>

(iii) $\|f_{\#}V\|B \leq \exp[-w \operatorname{diam} B]\|V\|B$,

<u>then</u> $f \in \underline{\underline{E}}(V,\sigma,w)$.

<u>Proof</u>: Since (ii) is the same as condition (a) in 4.9, we need only check condition (b). Let $\phi \in \mathscr{A}_w$. Then by (i), 4.2(1), and (iii),

$$\|f_{\#}V\|(\phi) - \|V\|(\phi)$$

$$= \int_B \phi(x)\,d\|f_{\#}V\|x - \int_B \phi(x)\,d\|V\|x$$

$$\le \sup\{\phi(x) : x \in B\}\|f_{\#}V\|B - \inf\{\phi(x) : x \in B\}\|V\|B$$

$$\le \exp[w \operatorname{diam} B]\inf\{\phi(x) : x \in B\}\|f_{\#}V\|B$$

$$- \inf\{\phi(x) : x \in B\}\|V\|B$$

$$\le 0. \qquad \Box$$

4.11. Approximation during small finite step.

Proposition: If $V \in \Omega$, $m \in \mathbb{N}$, $\phi \in \mathscr{A}_m$, and f_2 is as defined in 4.9, then

(i) $$|[\|f_{2\#}V\|(\phi) - \|V\|(\phi)]/\Delta t(m) - \delta(V,\phi)(h_{\varepsilon(m)}(V))| \le \varepsilon(m)^{41}\|V\|(\phi),$$

(ii) $$|\delta(V,\phi)(h_{\varepsilon(m)}(V)) - \delta(f_{2\#}V,\phi)(h_{\varepsilon(m)}(f_{2\#}V))| \le \varepsilon(m)^{16}\|V\|(\phi), \text{ and}$$

(iii) $$|\langle\Omega, |\Phi_{\varepsilon(m)}*\delta V|^2/\Phi_{\varepsilon(m)}*\|V\|\rangle - \langle\Omega, |\Phi_{\varepsilon(m)}*\delta f_{2\#}V|^2/\Phi_{\varepsilon(m)}*\|f_{2\#}V\|\rangle| < \varepsilon(m)^{15}\|V\|(\Omega).$$

__Proof__: Let $\varepsilon = \varepsilon(m)$ and $F = \Delta t(m)h_\varepsilon(V)$. Then 4.4 and 4.9 imply

(1) $$|F| < \varepsilon^{-4}\Delta t(m) < \varepsilon^{56}, \text{ and}$$

(2) $$\|DF\| < \varepsilon^{-8}\Delta t(m) < \varepsilon^{52}.$$

One may calculate, using (1), (2), 4.9(1), and 4.2(i,ii),

(3) $$\left||\Lambda_k Df_2 \circ S| - 1\right| < 2k\|DF\| < \varepsilon^{-9}\Delta t(m) < \varepsilon^{51},$$

$$\left||\Lambda_k Df_2 \circ S| - 1 - DF \cdot S\right| \leq 7k^2 k!^2 \|DF\|^2$$

$$\leq \varepsilon^{42}\Delta t(m),$$

$$|\phi(f_2(x)) - \phi(x)| \leq (\exp m|F(x)| - 1)\phi(x) < \varepsilon^{55}\phi(x),$$

$$|\phi(f_2(x)) - \phi(x) - F(x) \cdot D\phi(x)|$$

$$\leq [m^{-1}((\exp m|F(x)|) - 1) - |F(x)|]\phi(x)$$

$$< \varepsilon^{50}\Delta t(m)\phi(x).$$

Therefore, recalling 2.8 and 2.10(1),

$$|[\|f_{2\#}V\|(\phi) - \|V\|(\phi)]/\Delta t(m) - \delta(V,\phi)(h_\varepsilon(V))|$$

$$= |[\int \phi(f_2(x))|\Lambda_k Df_2(x)\circ S|dV(x,S) - \int \phi(x)dV(x,S)]/\Delta t(m)$$

$$- \int Dh_\varepsilon(V)(x)\cdot S\ \phi(x) + h_\varepsilon(V)(x)\cdot D\phi(x)dV(x,S)|$$

$$= \Delta t(m)^{-1} \int |[\phi(f_2(x)) - \phi(x)]|\Lambda_k Df_2(x)\circ S|$$

$$+ [|\Lambda_k Df_2(x)\circ S| - 1]\phi(x) - DF(x)\cdot S\ \phi(x)$$

$$- F(x)\cdot D\phi(x)|dV(x,S)$$

$$\leq \Delta t(m)^{-1} \int |[\phi(f_2(x)) - \phi(x)][|\Lambda_k Df_2(x)\circ S| - 1]$$

$$+ [\phi(f_2(x)) - \phi(x) - F(x)\cdot D\phi(x)]$$

$$+ [|\Lambda_k Df_2(x)\circ S| - 1 - DF(x)\cdot S]\phi(x)|\ dV(x,S)$$

$$< \int (\varepsilon^{46} + \varepsilon^{50} + \varepsilon^{42})\phi(x)d\|V\|x$$

$$< \varepsilon^{41}\|V\|(\phi),$$

which proves (i).

For (ii), one may calculate

$$\|Df_2(x)(S) - S\| \le [2\|DF(x)\|]^{1/2}$$

$$< 2\varepsilon^{26},$$

$$|D\Phi_\varepsilon(f_2(y) - x| < \varepsilon^{-4}\Phi_\varepsilon(f_2(y) - x) < \varepsilon^{-5}\Phi_\varepsilon(y - x),$$

$$|D\Phi_\varepsilon(f_2(y) - x) - D\Phi_\varepsilon(y - x)| < \varepsilon^{47}\Phi_\varepsilon(y - x).$$

Therefore, recalling 4.3(7) and (3),

(4) $$|\Phi_\varepsilon * \delta(f_{2\#}V)(x) - \Phi_\varepsilon * \delta V(x)|$$

$$= \left|\int T(D\Phi_\varepsilon(z - x))df_{2\#}V(z,T) - \int S(D\Phi_\varepsilon(y - x))dV(y,S)\right|$$

$$= \left|\int Df_2(y)(S)(D\Phi_\varepsilon(f_2(y) - x)|\Lambda_k Df_2{\circ}S|\right.$$

$$\left. - S(D\Phi_\varepsilon(y - x))dV(y,S)\right|$$

$$\le \int |D\Phi_\varepsilon(f_2(y) - x)|\left||\Lambda_k Df_2(y){\circ}S| - 1\right|$$

$$+ |Df_2(y)(S) - S||D\Phi_\varepsilon(f_2(f_2(y) - x)|$$

$$+ |D\Phi_\varepsilon(f_2(y) - x) - D\Phi_\varepsilon(y - x)|d\|V\|y$$

$$< \int \varepsilon^{-5}\Phi_\varepsilon(y - x)\varepsilon^{51} + 2\varepsilon^{26}\varepsilon^{-5}\Phi_\varepsilon(y - x)$$

$$+ \varepsilon^{47}\Phi_\varepsilon(y - x)d\|V\|y$$

$$< \varepsilon^{20}\Phi_\varepsilon * \|V\|(x).$$

Likewise,

(5) $$|\Phi_\varepsilon * \|f_{2\#}V\|(x) - \Phi_\varepsilon * \|V\|(x)|$$

$$= \left|\int \Phi_\varepsilon(f_2(y) - x)|\Lambda_k Df_2 \circ S| - \Phi_\varepsilon(y - x)\,d\|V\|y\right|$$

$$< \varepsilon^{51}\Phi_\varepsilon * \|V\|(x).$$

Hence

$$\left| \frac{\Phi_\varepsilon * \delta f_{2\#}V(x)}{\Phi_\varepsilon * \|f_{2\#}V\|(x)} - \frac{\Phi_\varepsilon * \delta V(x)}{\Phi_\varepsilon * \|V\|(x)} \right| < \varepsilon^{19},$$

and so

$$|h_\varepsilon(f_{2\#}V)(x) - h_\varepsilon(V)(x)| < \varepsilon^{19},$$

$$\|Dh_\varepsilon(f_{2\#}V)(x) - Dh_\varepsilon(V)(x)\| < \varepsilon^{15}.$$

Recalling 2.10(1) again,

$$\left| \delta(V,\phi)(h_\epsilon(V)) - \delta(f_{2\#}V)(h_\epsilon(f_{2\#}V)) \right|$$

$$= \left| \int Dh_\epsilon(V)(x) \cdot S\phi(x) + h_\epsilon(V)(x) \cdot D\phi(x)\, dV(x,S) \right.$$

$$\left. - \int Dh_\epsilon(f_{2\#}V)(y) \cdot T\phi(y) + h_\epsilon(f_{2\#}V)(y) \cdot D\phi(y)\, df_{2\#}V(y,T) \right|$$

$$\leq \int \left| Dh_\epsilon(V)(x) \cdot S\phi(x) - Dh_\epsilon(f_{2\#}V)(f_2(x)) \cdot (Df_2(x)(S))\phi(f_2(x)) \right.$$

$$\cdot \left| \Lambda_k Df_2(x) \circ S \right|$$

$$+ \left| h_\epsilon(V)(x) \cdot D\phi(x) - h_\epsilon(f_{2\#}V)(f_2(x)) \cdot D\phi(f_2(x)) \left| \Lambda_k Df_2(x) \circ S \right| \right|$$

$$dV(x,S)$$

$$\leq \epsilon^{16} \|V\|(\phi),$$

which proves (ii).

Conclusion (iii) follows from (4) and (5).   □

4.12. Constraints on motion.

Here we deduce upper bounds on the rate of change of the integral of a test function analogous to those of 3.4.

Proposition: If $V \in \underline{\Omega}$ and $m \in \underline{N}$, then

(i) $[\|f_{2\#}V\|(\Omega) - \|V\|(\Omega)]/\Delta t(m)$

$$\leq (-1 + m^{-4})\langle \Omega, |\Phi_{\varepsilon(m)} * \delta V|^2 / \Phi_{\varepsilon(m)} * \|V\|\rangle$$

$$+ (1 + m^{-4})\langle \Omega, |\Phi_{\varepsilon(m)} * \delta V|^2 / \Phi_{\varepsilon(m)} * \|V\|\rangle^{1/2} \|V\|(\Omega)^{1/2}$$

$$+ m^{-4}\|V\|(\Omega),$$

(ii) <u>if</u> $i \in \mathbf{N}$, $i < m$, <u>and</u> $\phi \in \mathscr{A}_i$, <u>then</u>

$$[\|f_{2\#}V\|(\phi) - \|V\|(\phi)]/\Delta t(m) \leq 2i^2\|V\|(\phi).$$

<u>Proof</u>: The proof of (i) will be a by-product of the proof of (ii).

Letting $\varepsilon = \varepsilon(m)$ and $h_\varepsilon = h_\varepsilon(V)$, and using 4.11(i), 2.10(2), 4.7(i), and 4.6(i,iii),

(1) $\{\|f_{2\#}V\|(\phi) - \|V\|(\phi)\}/\Delta t(m) \leq \delta(V,\phi)(h_\varepsilon) + \varepsilon^{41}\|V\|(\phi)$

$$\leq \delta V(\phi h_\varepsilon) + \int h_\varepsilon(x) \cdot S^\perp(D\phi(x))\, dV(x,S) + \varepsilon^{41}\|V\|(\phi)$$

$$\leq -\langle \phi, |\Phi_\varepsilon * \delta V|^2 / \Phi_\varepsilon * \|V\|\rangle - \langle \Phi_\varepsilon * \delta V, D\phi\rangle$$

$$+ 3\, ni\varepsilon^{1/2}\langle \phi, |\Phi_\varepsilon * \delta V|^2 / \Phi_\varepsilon * \|V\|\rangle$$

$$+ [2\, ni\varepsilon + ni\varepsilon^{-6}\, c_2(i,\varepsilon) + \varepsilon^{41}]\|V\|(\phi)$$

$$+ 2ni\varepsilon\langle \phi, |\Phi_\varepsilon * \delta V|^2 / \Phi_\varepsilon * \|V\|\rangle^{1/2}\|V\|(\phi)^{1/2}.$$

By 4.1(3), Schwarz' inequality, and 4.5(1),

(2) $$\langle|\Phi_\varepsilon*\delta V|,|D\phi|\rangle \le i\langle|\Phi_\varepsilon*\delta V|,\phi\rangle$$

$$\le i\langle|\Phi_\varepsilon*\delta V|^2/\Phi_\varepsilon*\|V\|,\phi\rangle^{1/2}\langle\Phi_\varepsilon*\|V\|,\phi\rangle^{1/2}$$

$$\le 2i\langle|\Phi_\varepsilon*\delta V|^2/\Phi_\varepsilon*\|V\|,\phi\rangle^{1/2}\|V\|(\phi)^{1/2}.$$

Hence

(3) $$\{\|f_{2\#}V\|(\phi) - \|V\|(\phi)\}/\Delta t(m)$$

$$\le [-1 + 3\ ni\varepsilon^{1/2}]\langle\phi,|\Phi_\varepsilon*\delta V|^2/\Phi_\varepsilon*\|V\|\rangle$$

$$+ [2i + 2\ ni\varepsilon]\langle\phi,|\Phi_\varepsilon*\delta V|^2/\Phi_\varepsilon*\|V\|\rangle^{1/2}\|V\|(\phi)^{1/2}$$

$$+ [2\ ni\varepsilon + ni\varepsilon^{-6}\ c_2(i,\varepsilon) + \varepsilon^{41}]\|V\|(\phi),$$

which, by the properties of $\varepsilon(m)$ in 4.9(1), is less than

(4) $$[-1 + m^{-4}]\langle\phi,|\Phi_\varepsilon*\delta V|^2/\Phi_\varepsilon*\|V\|\rangle$$

$$+ [2i + m^{-4}]\langle\phi,|\Phi_\varepsilon*\delta V|^2/\Phi_\varepsilon*\|V\|\rangle^{1/2}\|V\|(\phi)^{1/]} + m^{-4}\|V\|(\phi).$$

Taking $\phi = \Omega$ and $i = 1$ gives (i). The maximum value of expression (4) is

(5) $$((1/4)[2i + m^{-4}]^2[1 - m^{-4}]^{-1} + m^{-4})\|V\|(\phi),$$

which proves (ii). □

4.13. Towards a varifold moving by its mean curvature

In this section we do the construction that will give us a varifold moving by its mean curvature. The rest of the chapter is devoted to showing that we do indeed have a solution with the claimed properties.

Let $V_0 \in \underline{\underline{\Omega}}$. For all positive integers m and p, choose the varifolds $V^*_{m,p\Delta t(m)}$, $V_{m,p\Delta t(m)} \in \underline{\underline{\Omega}}$ inductively as follows:

(1) $$V_{m,0} = V_0,$$

(2) $$V^*_{m,(p+1)\Delta t(m)} = f_{1\#}V_{m,p\Delta t(m)},$$

$$V_{m,(p+1)\Delta t(m)} = f_{2\#}V^*_{m,(p+1)\Delta t(m)},$$

where $f_1 \in E(V_{m,p\Delta t(m)}, \sigma(m), m)$ is chosen so that

$$\|f_{1\#}V\|(\Omega) - \|V\|(\Omega) \leq (1 - m^{-5})\Delta_{\sigma(m),m}\|V\|(\Omega)$$

and

$$f_2(x) = x + \Delta t(m)h_{\varepsilon(m)}(V^*_{m,(p+1)\Delta t(m)}).$$

Let $\underline{\underline{Q2}}$ denote the set of nonnegative dyadic rationals, and let

$$\mathbf{Q}_m = \{p\Delta t(m) : p \in N\}.$$

It follows from 4.12(ii) that for fixed $t \in \mathbf{Q2}$ we have for large enough m

$$\|V_{m,t}\|(\Omega) \le e^{2t}\|V_0\|(\Omega).$$

By 4.1, the set

$$\{V \in \Omega\mathbf{V}: \|V\|(\Omega) \le e^{2t}\|V_0\|(\Omega)\}$$

is compact in the Ω topology, so we may use a Cantor diagonal process to choose a subsequence m_i, $i \in N$, such that $\Omega \lim_{i\to\infty} V_{m_i,t}$ exists for each $t \in \mathbf{Q2}$. Without loss of generality, we may assume $\Omega \lim_{m\to\infty} V_{m,t} = V_t$ for each $t \in \mathbf{Q2}$.

The reason for including $V^*_{m,(p+1)\Delta t(m)}$ explicitly instead of defining $V_{m,(p+1)\Delta t(m)}$ directly from $V_{m,p\Delta t(m)}$ is that later we will need to talk about smoothed mean curvature and Lipschitz deformations of the same varifold, $V_{m,t}$. That does not fit in with the alternating nature of the procedure just defined, but by 4.11 the properties of the smoothed mean curvature of $V_{m,t}$ are well approximated by the properties of the smoothed mean curvature of $V^*_{m,t}$. It follows from 4.12(ii) that for $t \in \mathbf{Q2}$ we have

$$\Omega \lim_{m\to\infty} \|V^*_{m,t}\| = \|V_t\|.$$

4.14. Continuity of $\|V_t\|$.

Proposition: If $V_0 \in \Omega$ and V_t is as defined in 4.13, then

(a) We may extend the domain of definition of $\|V_t\|$ to all $t > 0$, $t \notin \underline{Q2}$, by defining measures on $\underline{R}^n$

$$\|V_t\| = \Omega \lim_{\substack{s \to t^- \\ s \in \underline{Q2}}} \|V_s\|.$$

(b) If $i \in \underline{N}$ and $\phi \in \mathscr{A}_i$ then for all $t > 0$

$$\bar{D}\|V_t\|(\phi) \leq 2i^2 \|V_t\|(\phi).$$

(c) $\|V_t\|(\phi)$ is a continuous function of t at almost every $t > 0$.

(d) If $t_0 > 0$ and $\|V_t\|(\phi)$ is discontinuous at t_0, then $\|V_t\|(\phi)$ has a jump decrease at t_0.

(e) For any $t > 0$, $\Omega \lim_{t \to t_0^-} \|V_t\|$ exists and

$$\Omega \lim_{t \to t_0^-} \|V_t\| \geq \Omega \lim_{t \to t_0^+} \|V_t\|.$$

(f) $\|V_t\|$ is a continuous function of t at almost all $t \geq 0$.

(g) <u>If</u> $\|V_t\|$ <u>is continuous at</u> t_0 <u>and</u> $s_1, S_2, \ldots, m_1, m_2, \ldots$ <u>are sequence with</u> $s_i \in Q_{m_i}$ <u>and</u> $\lim_{m\to\infty} s_m = t_0$, <u>then</u>

$$\|V_{t_0}\| = \Omega \lim_{i\to\infty} \|V_{m_i, s_i}\|.$$

<u>Remark</u>: The full definition of V_t for all t will have to wait until we can show rectifiability.

<u>Proof</u>: If $i \in \underline{\underline{N}}$, $\phi \in \mathscr{A}_i$, $m \in \underline{\underline{N}}$, $m > i$, and $r \in \underline{\underline{Q}}_m$ then from 4.12(ii) we have

(1) $$[\|V_{m,r+\Delta t(m)}\|(\phi) - \|V_{m,r}\|(\phi)]/\Delta t(m) \leq 2i^2\|V_{m,r}\|(\phi),$$

which implies

(2) $$\|V_s\|(\phi) \leq \exp[2i^2|s - r|]\|V_r\|(\phi)$$

for all $r, s \in Q2$ with $r < s$. Therefore for any $t > 0$

(3) $$\lim_{\substack{s\to t^- \\ s\in\underline{\underline{Q2}}}} \|V_s\|(\phi)$$

exists. Since the set of Radon measures μ on $\underline{\underline{R}}^n$ with $\mu(\Omega)$ bounded is compact in the Ω topology, (3) says that we may define

$$\|V_t\| = \lim_{\substack{s\to t^- \\ s\in\underline{\underline{Q2}}}} \|V_s\|$$

for $t \notin \underline{Q2}$.

Now (b) follows from (2), and (c) and (d) follow from (b). Since a test function $\psi \in \underline{\Omega C}(\underline{R}^n)$ can be approximated by $\phi \in \mathscr{A}_i$ for large enough i, (2) implies

$$(5) \qquad \lim_{t \to t_0^-} \|V_t\|(\psi) \geq \lim_{t \to t_0^+} \|V_t\|(\psi)$$

which proves (e). Furthermore, $\|V_t\|(\psi)$ can have only a countable number of discontinuities, and since the space of test functions is separable, $\|V_t\|$ is a continuous function of t at almost all $t \geq 0$, which is (f). Whenever $\|V_t\|$ is continuous at t_0 conclusion (g) follows from (1) and the definition of $\|V_t\|$ in 4.13. □

4.15. Agreement on smooth manifolds.

If the initial varifold represents a smooth manifold, then it is clearly desirable that the approximation procedure described in this chapter should agree with the more straightforward mapping approach described in 3.1, at least as long as the latter works. Since the smoothed mean curvature would be very near the mean curvature in such a case, we could say the two approaches agree if we can show that the only eligible Lipschitz maps f_1 would leave the varifold fixed.

Theorem: Suppose $0 < \gamma < 1$. Then there is $m_0 \in \underline{N}$ such that if $m > m_0$, if $M \in \underline{\Omega}$ represents a k-dimensional manifold of class $\underline{C}^3$ without boundary embedded in $\underline{R}^n$ with a normal neighborhood of radius $\gamma/5$ and with all sectional curvatures of magnitude less than $1/\gamma$, and if $f \in \underline{E}(M, \sigma(m), m)$ then

$$f_{\#}M = M.$$

Proof: The theorem will follow if we can show that for large enough m there is $\phi \in \mathcal{A}_m$ such that $f_{\#}M \neq M$ implies $\|f_{\#}M\|(\phi) > \|M\|(\phi)$.

Let N be a normal neighborhood of $\mathrm{spt}\|M\|$ of radius $\gamma/5$ and let $\pi: N \to \mathrm{spt}\|M\|$ be the nearest point retraction. Since $\mathrm{spt}\|M\|$ is a manifold without boundary, we have $\pi \circ f(\mathrm{spt}\|M\|) = \mathrm{spt}\|M\|$ and for any nonnegative continuous function ψ

$$\|\pi_{\#}f_{\#}M\|(\psi) \geq \|M\|(\psi).$$

Therefore it is sufficient to find conditions on ϕ that will guarantee

$$(1) \qquad \|f_{\#}M\|(\phi) \geq \|\pi_{\#}f_{\#}M\|(\phi),$$

if $f_{\#}M \neq \pi_{\#}f_{\#}M$. We cannot take $\phi(x) = \mathrm{dist}(x,M)^2$ because $\phi(x) \neq 0$ by 4.2(i). From 2.8 we have

(2) $$\|\pi_{\#} f_{\#} M\|(\phi) = \int |\Lambda_k D\pi(x) \circ S| \phi(\pi(x)) \, d f_{\#} M(x,S).$$

To calculate $|\Lambda_k D\pi(x) \circ S|$, suppose that $\pi(x) = 0$, $\operatorname{spt}\|M\|$ is the graph of $F: \underline{\underline{R}}^k \to \underline{\underline{R}}^{n-k}$ in a neighborhood of 0, $F(0) = 0$, $\operatorname{Tan}^k(\|V\|, 0) = \underline{\underline{e}}_1 \wedge \ldots \wedge \underline{\underline{e}}_k$, and x is on the x_{k+1} axis.

We may represent $S \in \underline{\underline{G}}(n,k)$ by

$$S = \sum_{\lambda \in \Lambda(n,)} \alpha_\lambda \underline{\underline{e}}_{\lambda_1} \wedge \ldots \wedge \underline{\underline{e}}_{\lambda_k}$$

with $\sum \alpha_\lambda^a = 1$ (see [FH 1.3.2]). Thus

$$\Lambda_k D\pi(x) \circ S = \sum_\lambda \alpha_\lambda \, D\pi(x)(\underline{\underline{e}}_{\lambda_1}) \wedge \ldots \wedge D\pi(x)(\underline{\underline{e}}_{\lambda_k}).$$

Clearly $D\pi(x)(\underline{\underline{e}}_j) = 0$ for $j > k$, and for $1 \leq j \leq k$ calculation shows that

$$D\pi(x)(\underline{\underline{e}}_j) = \underline{\underline{e}}_j + x_{k+1} \sum_{i=1}^{k} (\partial^2 F_{k+1}(0)/\partial x_j \partial x_i) \underline{\underline{e}}_i.$$

Therefore, one may compute

$$|\Lambda_k D\pi(x) \circ S| \leq |D\pi(x)(\underline{\underline{e}}_1) \wedge \ldots \wedge D\pi(x)(\underline{\underline{e}}_k)|$$

$$\leq 1 + x_{k+1} \sum_{j=1}^{k} \partial^2 F_{k+1}(0)/\partial x_j^2 + (k+1)!\gamma^{-1} x_{k+1}^2.$$

From differential geometry, we have

$$\underline{h}(M,0) = \sum_{j=k+1}^{n} \underline{\underline{e}}_j \sum_{i=1}^{k} \partial^2 F_j(0)/x_i^2,$$

so in general we have

$$|\Lambda_k D\pi(x)\circ S| \le 1 + (x - \pi(x))\cdot \underline{\underline{h}}(M,\pi(x)) + (k + 1)!\gamma^{-1}|x - \pi(x)|^2.$$

It now follows from (2) that if $\phi \in \mathscr{A}_m$ and

(3) $$\|D^2\phi(y) - D^2\phi(\pi(y))\| < \gamma^{-1}|y - \pi(y)|\phi(\pi(y)) \quad \text{for} \quad y \in N,$$

then, using Taylor's formula on $\phi(\pi(x))$,

(4) $$\|\pi_\# f_\# M\|(\phi) \le \int \Big[1 + (x - \pi(x))\cdot \underline{\underline{h}}(M,\pi(x)) + (k + 1)!\gamma^{-1}|x - \pi(x)|^2\Big]\Big[\phi(x) - D\phi(\pi(x))(x - \pi(x)) - D^2\phi(\pi(x))(\pi(x) - x,\pi(x) - x)/2 + \gamma^{-1}|x - \pi(x)|^3\phi(\pi(x))\Big]\, d\|f_\# M\|x.$$

If we require that

(5) $$D\phi(z) = \phi(z)\underline{\underline{h}}(M,z) \quad \text{and}$$

(6) $$D^2\phi(z)(w,w) \ge 4(k + 1)!\gamma^{-1}$$

for $z \in M$ and w normal to M at z, then (4) becomes

$$\|\pi_{\#}f_{\#}M\|(\phi) \leq \|f_{\#}M\|(\phi) + \int - D^2\phi(\pi(x))(\pi(x) - x, \pi(x) - x)/2$$

$$+ (k + 1)!\gamma^{-1}\phi(\pi(x))|\pi(x) - x|^2 + \gamma^{-1}|x - \pi(x)|^3\phi(\pi(x))d\|f_{\#}M\|x$$

$$\leq \|f_{\#}M\|(\phi) - \int \phi(\pi(x))|\pi(x) - x|^2\gamma^{-1}d\|f_{\#}M\|x.$$

Clearly, for m depending on γ and k, there exists $\phi \in \mathscr{A}_m$ satisfying (3), (5), and (6). Thus (1) holds unless $\pi(x) = x$ $\|f_{\#}M\|$ almost everywhere. □

4.16. Towards rectifiability.

The next few sections show that if a sequence of varifolds in $\underline{\underline{\Omega}}$ have bounded rates of mass loss, then their limit varifold will be rectifiable. The main tasks are to prove a lower density bound and that the limit has bounded first variation, since these conditions by [AW1 5.5(1)] imply rectifiability. This first proposition states that for small balls of low density there are Lipschitz maps reducing mass drastically.

Proposition: *There is a constant* $c_3 > 0$ *such that if* $V \in \underline{\underline{\Omega}}$, $0 \in \operatorname{spt}\|V\|$, *and* $\|V\|\underline{\underline{B}}(0,1) < c_3$ *then there exist* $0 < R < 1$ *and a Lipschitz map* $f\colon \underline{\underline{R}}^n \to \underline{\underline{R}}^n$ *such that*

i) $f(x) = x$ *for* $x \notin \underline{\underline{B}}(0,R)$,

ii) $f(x) \in \underline{\underline{B}}(0,R)$ *for* $x \in \underline{\underline{B}}(0,R)$, *and*

iii) $\|f_{\#}V\|\underline{\underline{B}}(0,R) \leq (1/2)\|V\|\underline{\underline{B}}(0,R)$.

Proof: For $r > 0$ let $\mu(r) = \|V\|\mathbb{B}(0,r)$. By [AF1.9(2)], for almost all $r > 0$ there exists a Lipschitz map $f_r: \mathbb{R}^n \to \mathbb{R}^n$ satisfying conclusions (i) and (ii) such that

$$\|f_{r\#}V\|\mathbb{B}(0,r) \le 2[2n^{2k}\mu'(r)]^{k/(k-1)}.$$

If conclusion (iii) were false, then for almost all $0 < r < 1$

(1) $$2[2n^{2k}\mu'(r)]^{k/(k-1)} \ge \mu(r)/2.$$

Since $0 \in \operatorname{spt}\|V\|$, we have $\mu(r) > 0$ for $r > 0$, so (1) may be integrated to

$$\mu(1) \ge [2n^{2k}]^{-k^2(k-1)}4^{-k}.$$

Thus we need only choose $c_3 = 4^{-k}[2n^{2k}]^{-k^2/(k-1)}$. □

4.17. *Monotonicity*.

This lemma [AW1 5.1(3)] says that the rate of decrease of density ratios as a function of radius is limited by the amount of curvature present.

Lemma: *Suppose* $V \in \mathbf{V}_k(\mathbb{R}^n)$, $0 \le M < \infty$, $0 < R_1 < R_2 < \infty$, $a \in \mathbb{R}^n$, *and*

$$\|\delta V\|\mathbb{B}(a,r) \le M\|V\|\mathbb{B}(a,r)$$

whenever $R_1 < r < R_2$. *Then*

$$(\exp Mr) r^{-k} \|V\| \underline{B}(a,r)$$

is nondecreasing in r *for* $R_1 < r < R_2$. □

4.18. *Curvature of limit.*

Proposition: *If* $V_1, V_2, \ldots \in \underline{\Omega}$, $\Omega \lim_{m\to\infty} V_m = V \in \underline{\Omega V}$, $i \in \underline{N}$, *and* $\phi \in \mathscr{A}_i$ *then*

$$\int |\underline{h}(V,x)|^2 \phi(x)\, d\|V\|x$$

$$\leq \liminf_{m\to\infty} \langle \phi, |\Phi_{\varepsilon(m)} * \delta V_m|^2 / \Phi_{\varepsilon(m)} * \|V_m\| \rangle.$$

Proof: Suppose $C > 0$ and

$$\int |\underline{h}(V,x)|^2 \phi(x)\, d\|V\|x > C^2.$$

Then there exists a smooth $g\colon \underline{R}^n \to \underline{R}^n$ such that $\sup\{|D(\phi g)(x)|/\Omega(x) : x \in R^n\} < \infty$,

$$\int \underline{h}(V,x) \cdot g(x)\phi(x)\, d\|V\|x = \delta V(\phi g) > C^2, \quad \text{and}$$

$$\int |g(x)|^2 \phi(x)\, d\|V\|x < C^2.$$

By 4.1 and 4.3(10) we have

$$\delta V(\phi g) = \lim_{m\to\infty} \Phi_{\varepsilon(m)} * \delta V_m(\phi g),$$

hence, using Schwarz' inequality,

$$C^2 < \lim_{m\to\infty} \Phi_{\varepsilon(m)} * \delta V_m(\phi g)$$

$$\leq \liminf_{m\to\infty} \langle |\Phi_{\varepsilon(m)} * \delta V_m|^2/\Phi_{\varepsilon(m)} * \|V_m\|, \phi\rangle^{1/2} \cdot \langle \Phi_{\varepsilon(m)} * \|V_m\|, \phi g^2\rangle^{1/2}$$

$$\leq C \liminf_{m\to\infty} \langle |\Phi_{\varepsilon(m)} * \delta V_m|^2/\Phi_{\varepsilon(m)} * \|V_m\|, \phi\rangle^{1/2}.$$

Thus

$$C^2 \leq \liminf_{m\to\infty} \langle \phi, |\Phi_{\varepsilon(m)} * \delta V_m|^2/\Phi_{\varepsilon(m)} * \|V\|\rangle$$

and the conclusion follows. □

Remark: The possibility of inequality in this proposition is good evidence that we do not want to require equality in 3.3(1).

4.19. Rectifiability.

Now we show that if a sequence of varifolds has a common bound on the rate of mass loss, then a limit varifold is rectifiable. This is not yet talking about V_t being rectifiable. That will be discussed after we prove the corresponding result an integrality.

Theorem: *If* $B > 0$, $V_1, V_2, \ldots \in \underline{\Omega}$, $\Omega \lim_{m\to\infty} V_m = V \in \underline{\Omega V}$, $\|V_m\|(\Omega) < B$ *and*

$$(1) \qquad \langle \Omega, |\Phi_{\varepsilon(m)} * \delta V_m|^2 / \Phi_{\varepsilon(m)} * \|V_m\|\rangle - \Delta_{\sigma(m),m}\|V_m\|(\Omega)/\Delta t(m) < B$$

for all $m \in \underline{N}$, *then*

(2) $\Theta^{*k}(\|V\|, x) \geq c_3/16\underline{\alpha}$ *for* $\|V\|$ *almost all* $x \in \underline{R}^n$, *and*

(3) V *is rectifiable.*

Proof: For $0 < R < (4B)^{-1}$ define

$$F_R = \{x \in \underline{R}^n \colon R^{-k}\|V\|\underline{B}(x,R) < c_3/16\}.$$

If $|x - y| < (1 - 2^{-1/k})R$, then

$$\underline{B}(y, 2^{-1/k}R) \subset \underline{U}(x,R),$$

and so

$$(2^{-1/k}R)^{-k}\|V\|\underline{B}(y, 2^{-1/k}R) \geq 2^{-1}R^{-k}\|V\|\underline{U}(x,R).$$

Since $\Omega \lim_{m\to\infty} V_m = V$ and $\Omega \lim_{m\to\infty} \Phi_{\varepsilon(m)} * V_m = V$ by 4.3(10), for for each $x \in F_R$ there must be $M(x) \in \underline{N}$ such that $m > M(x)$ implies that for $|x - y| < (1 - 2^{-1/k})R$ we have

(4) $$(2^{-1/k}R)^{-k}\|\Phi_{\varepsilon(m)}*V_m\|\underline{\underline{B}}(y,2^{-1/k}R) < c_3/8.$$

Choose $M_1 \in \underline{\underline{N}}$ such that

$$\|V\|(\Omega|\{x \in F_R: M(x) \leq M_1\}) > (1/2)\|V\|(\Omega|F_R).$$

Define

$$G_R = \{y \in \underline{\underline{R}}^n: \mathrm{dist}(y,\{x \in F_R: M(x) \leq M_1\}) < (1 - 2^{-1/k})R\}.$$

Since G_R is open, there is an $M_2 > M_1$ such that if $m > M_2$ then

(5) $$\|V_m\|(\Omega|G_R) > (1/2)\|V\|(\Omega|G_R) > (1/4)\|V\|(\Omega|F_R).$$

Choose $M_3 > M_2$ such that $\sigma(M_3) < R/2$, and let $m > M_3$. Define

(6) $$E_1(R,m) = \{x \in G_R \cap \mathrm{spt}\|V_m\|: \Theta^k(\|V_m\|,x) \geq 1$$
$$\text{and } \sigma(m)^{-k}\|\Phi_{\varepsilon(m)}*V_m\|\underline{\underline{B}}(x,\sigma(m)) > c_3/4\}$$

and

(7) $$E_2(R,m) = \{x \in G_R \cap \mathrm{spt}\|V_m\|: \Theta^k(\|V_m\|,x) \geq 1$$
$$\text{and } \sigma(m)^{-k}\|\Phi_{\varepsilon(m)}*V_m\|\underline{\underline{B}}(x,\sigma(m)) \leq c_3/4\}.$$

By the definition of $\underline{\underline{\Omega}}$, we have $\Theta^k(\|V_m\|,x) \geq 1$ for $\|V_m\|$

almost all $x \in \mathbb{R}^n$, so

(8) $$\|V_m\|(\Omega|E_1(R,m) \cup E_2(R,m)) = \|V_m\|(\Omega|G_R).$$

Suppose $x \in E_1(R,m)$. It follows from (4), (6), and 4.17 that there is $\sigma(m) < r(x) < 2^{-1/k}R$ such that

$$\|\delta\Phi_{\varepsilon(m)}*V_m\|\mathbb{B}(x,r(x))$$

$$\geq (2^{-1/k}R - \sigma(m))^{-1}(\ln 2)\|\Phi_{\varepsilon(m)}*V_m\|\mathbb{B}(x,r(x))$$

$$\geq (1/2R)\|\Phi_{\varepsilon(m)}*V_m\|\mathbb{B}(x,r(x)).$$

Since $\delta\Phi_{\varepsilon(m)}*V_m = \Phi_{\varepsilon(m)}*\delta V_m$ by 4.3 and $\Omega(y) < \Omega(z)\exp|y - z|$ by 4.2(i), we have

(9) $$\|\Phi_{\varepsilon(m)}*\delta V_m\|(\Omega|\mathbb{B}(x,r(x))$$

$$\geq (\exp -2R)(1/2R)\|\Phi_{\varepsilon(m)}*V_m\|(\Omega|\mathbb{B}(x,r(x))).$$

By the Besicovitch covering theorem 2.2, we may choose a family of disjoint balls $\mathbb{B}(x,r(x))$ such that, if we denote their union by W, then

(10) $$\|\Phi_{\varepsilon(m)}*V_m\|(\Omega|W)$$

$$\geq \mathbb{B}(n)^{-1}\|\Phi_{\varepsilon(m)}*V_m\|(\Omega| \cup\{\mathbb{B}(x,r(x)): x \in E_1(R,m)\}).$$

Since $r(x) > \sigma(m)$, it follows from the definition of $\varepsilon(m)$ that

$$\text{(11)} \qquad \|\Phi_{\varepsilon(m)} * V_m\|(\Omega| \cup \{\underline{\underline{B}}(x,r(x)) : x \in E_1(R,m)\})$$

$$\geq (1/2)\|V_m\|E_1(R,m).$$

Putting (9), (10), and (11) together gives

$$\|\Phi_{\varepsilon(m)} * \delta V_m\|(\Omega) \geq \underline{\underline{B}}(n)^{-1}(\exp -2R)(1/4R)\|V_m\|E_1(R,m)$$

or, by Schwarz' inequality,

$$\langle \Omega, |\Phi_{\varepsilon(m)} * \delta V_m|^2/\Phi_{\varepsilon(m)} * \|V_m\|\rangle \|\Phi_{\varepsilon(m)} * V_m\|(\Omega)$$

$$\geq [\underline{\underline{B}}(n)^{-1}(\exp -2R)(1/4R)\|V_m\|E_1(R,m)]^2.$$

Hypothesis (1) now implies that

$$\text{(12)} \qquad \limsup_{m\to\infty} \|V_m\|E_1(R,m) \leq \underline{\underline{B}}(n)(\exp 2R)4R(B\|V\|(\Omega))^{1/2}.$$

Now suppose that $x \in E_2(R,m)$. It follows from the definition of $\varepsilon(m)$ that

$$\|V_m\|\underline{\underline{B}}(x, 2^{-1/k}\sigma(m)) \leq 2\|\Phi_{\varepsilon(m)} * V_m\|\underline{\underline{B}}(x,\sigma(m)),$$

so (7) implies that

$$(2^{-1/k}\sigma(m))^{-k}\|V_m\|\underline{\underline{B}}(x,2^{-1/k}\sigma(m)) \le c_3.$$

It follows from 4.16 that there exists $0 < r(x) < 2^{-1/k}\sigma(m)$ and a Lipschitz map $f_x: \underline{\underline{R}}^n \to \underline{\underline{R}}^n$ such that

$$f_x(y) = y \quad \text{for} \quad y \notin \underline{\underline{B}}(x,r(x)),$$

$$f_x(y) \in \underline{\underline{B}}(x,r(x)) \quad \text{for} \quad y \in \underline{\underline{B}}(x,r(x)), \quad \text{and}$$

$$\|f_{x\#}V_m\|\underline{\underline{B}}(x,r(x)) \le (1/2)\|V_m\|\underline{\underline{B}}(x,r(x)).$$

By the properties of Ω and $\sigma(m)$,

(13) $$\|f_{x\#}V_m\|(\Omega|\underline{\underline{B}}(x,r(x)) \le 2^{-1/2}\|V_m\|(\Omega|\underline{\underline{B}}(x,r(x))).$$

By the Besicovitch covering theorem, we may choose a subset $\{x_\lambda: \lambda \in \Lambda\} \subset E_2(R,m)$ such that all the $\underline{\underline{B}}(x_\lambda,r(x_\lambda))$ are disjoint and

(14) $$\|V_m\|(\Omega|\cup\{\underline{\underline{B}}(x_\lambda,r(x_\lambda)): \lambda \in \Lambda\})$$

$$\ge \underline{\underline{B}}(n)^{-1}\|V_m\|(\Omega|\cup\{\underline{\underline{B}}(x,r(x)): x \in E_2(R,m)\})$$

$$\ge \underline{\underline{B}}(n)^{-1}\|V_m\|(\Omega|E_2(R,m)).$$

Define the Lipschitz map $f: \underline{\underline{R}}^n \to \underline{\underline{R}}^n$ by

$$f(y) = \begin{cases} f_{x_\lambda}(y) & \text{if} \quad f_{x_\lambda}(y) \neq y \quad \text{for some} \quad \lambda \in \Lambda \\ y & \text{otherwise.} \end{cases}$$

By (13) and 4.10, we have $f \in \underline{\underline{E}}(V_m, \sigma(m), m)$. Therefore, by (13) and (14),

$$-\Delta_{\sigma(m),m}\|V_m\|(\Omega) \geq (1 - 2^{-1/2})\underline{\underline{B}}(n)^{-1}\|V_m\|(\Omega|E_2(R,m)).$$

Hypothesis (1) and $\Delta t(m) \to 0$ imply that

(15) $$\limsup_{m\to\infty} \|V_m\|(\Omega|E_2(R,m)) = 0.$$

Combining (12), (15), (8), and (5) yields

$$\|V\|(\Omega|F_R) \leq 4\underline{\underline{B}}(n)(\exp 2R)4RB.$$

This implies

$$\lim_{R\to 0} \|V\|(\Omega|F_R) = 0.$$

Hence

$$\Theta^{*k}(\|V\|,x) \geq c_3/16\underline{\underline{\alpha}}$$

for $\|V\|$ almost all $x \in \underline{\underline{R}}^n$, which proves (2).

By 4.18 and hypothesis (1), we have

$$\int |\underline{\underline{h}}(V,x)|^2 \Omega(x)\, d\|V\|x \leq B.$$

Hence $\|\delta V\|$ is a Radon measure, and we may apply [AW1 5.5(1)] to conclude that $V \in \underline{\underline{RV}}_k(\underline{\underline{R}}^n)$, which proves (3). □

4.20. Towards integrality.

A sequence of integral varifolds will converge to an integral varifold under the same hypotheses as we had for rectifiable varifolds in 4.19. To prove this is the purpose of the next several sections. The proof follows the same ideas as the proof of the compactness theorem for integral varifolds in [AW1 6.4], but is necessarily more complex. We want to show that the densities of the limit are integers. Knowing already that the limit varifold is rectifiable, we show that non-integral density ratios in the approximation varifolds come from "holes" and therefore lead to large rates of mass loss. This first lemma is analagous to 4.16 and handles holes too small for the smoothed mean curvature to detect.

Lemma: *If* $\nu \in \underline{\underline{N}}$, $0 < \mu < 1$, *and* $0 < \zeta < 1$ *then there is* $\gamma > 0$ *such that if*

(1) $V \in \underline{\underline{IV}}_k(\underline{\underline{R}}^n)$, $\sigma > 0$, $w > 0$, $0 < R < \sigma$, $0 < 2\rho < \sigma$,

$$(1 - \zeta)/2\nu > 1 - \exp[-4w\sigma], \quad \rho/R > \mu$$

(2) $$T = \underline{\underline{e}}_1 \wedge \ldots \wedge \underline{\underline{e}}_k \in \underline{\underline{G}}(n,k);$$

(3) $$Y \subset T^{\perp}, \quad \text{diam } Y < \sigma, \quad \text{and} \quad \nu = \sum \{\Theta^k(\|V\|,y): y \in Y\};$$

(4) for $r > 0$ and $\xi > 0$ we define

$$E(r,\xi) = \{x \in \underline{\underline{R}}^n: |T(x)| \leq r, \text{dist}(T^{\perp}(x),Y) < \xi\};$$

(5) if $0 < r \leq R$ then

$$\int_{E(r,2\rho)} \|S - T\| dV(x,S) < \gamma \underline{\underline{\alpha}} r^k; \quad \text{and}$$

(6) if $0 < r \leq R$ then

$$\Delta_{\sigma,w} \|V\| E(r,\rho) > -\gamma \underline{\underline{\alpha}} r^k;$$

then

(7) $$\|V\| E(R,\rho) \geq (\nu - \zeta) \underline{\underline{\alpha}} R^k.$$

Proof: Define

(8) $$r = \inf\{s > 0: \|V\| E(s,(1 + s/R)\rho) \leq (\nu - \zeta) \underline{\underline{\alpha}} s^k\}.$$

Hypothesis (3) guarantees that $r > 0$. If $r \geq R$, then we are done. Otherwise, we look for a contradiction to (6). Letting $\rho_1 = (1 + r/R)\rho$, the properties of Radon measures imply

$$\|V\| E(r,\rho_1) = (\nu - \zeta) \underline{\underline{\alpha}} r^k.$$

For the rest of this proof, we shall suppose that $V = V \llcorner E(r,\rho_1)$. Noting that $T_\# V$ is an integral varifold, we define the set of "holes" $A_0 \subset \underline{\underline{R}}^k$ to consist of all $a \in \underline{\underline{U}}^k(0,r)$ such that

$$\Theta^k(\|T_\# V\|,a) \leq \nu - 1.$$

Since

$$(\nu - \zeta)\underline{\underline{\alpha}}r^k \geq \nu(\underline{\underline{\alpha}}r^k - \mathscr{H}^k(A_0)),$$

we have

$$\mathscr{H}^k(A_0) \geq \zeta\nu^{-1}\underline{\underline{\alpha}}r^k.$$

Let $0 < \xi < \rho r/R$ and $\eta > 0$ be arbitrary. By the definitions of induced mapping and density, there are $\delta > 0$ and $A \subset A_0$ such that

(9) $$\mathscr{H}^k(A) \geq (1 - \eta)\zeta\nu^{-1}\underline{\underline{\alpha}}r^k$$

and for each $a \in A$ we have $|a| + \delta < r$,

(10) $$\int_{\underline{\underline{C}}(T,a,\delta)} |\Lambda_k DT \circ D|\,dV(x,S) < (\nu - 1 + \eta)\underline{\underline{\alpha}}\delta^k, \quad \text{and}$$

(11) $$\|V\|\underline{\underline{C}}(T,a,\delta) < \eta\underline{\underline{\alpha}}\delta^{k-1}.$$

For each $a \in A$ we will now construct a Lipschitz map $f(a): \mathbb{R}^n \to \mathbb{R}^n$ that essentially expands a hole to fill up $E(r,\rho_1)$, replacing V by a varifold whose mass we can estimate by (10) and (11). Define $a^* = (1 - \delta/r)a$,

$$E_1(a) = \{x \in \mathbb{R}^n: |T(x) - a^*| \leq 2\delta\xi^{-1}(\rho_1 - \operatorname{dist}(T^{\perp}(x),Y)),$$

$$|T(x) - a| < \delta, \text{ and } \operatorname{dist}(T^{\perp}(x),Y) < \rho_1\},$$

$$E_2(a) = \{x \in R^n: |T(x) - a^*| \leq 2r\xi^{-1}(\rho_1 - \operatorname{dist}(T^{\perp}(x),Y)),$$

$$|T(x)| \leq r, \ \operatorname{dist}(T^{\perp}(x),Y) < \rho_1\} \sim E_1(a),$$

$$E_3(a) = \{x \in E_2(a): \operatorname{dist}(T^{\perp}(x),Y) < \rho_1 - \xi\}, \text{ and}$$

$$E_4(a) = E_2(a) \sim E_3(a).$$

Let $f(a): \mathbb{R}^n \to \mathbb{R}^n$ be the Lipschitz map which leaves $\mathbb{R}^n \sim (E_1(a) \cup E_2(a))$ fixed, projects $E_2(a)$ radially from $\{a^*\} \times \mathbb{R}^{n-k}$ to $\partial(E_1(a) \cup E_2(a))$, and expands $E_1(a)$ radially from $\{a^*\} \times \mathbb{R}^{n-k}$ by a factor of $r/\delta(a)$.

Next we calculate the mass of $f(a)_{\#}V$. At each $x \in \mathbb{R}^n$ define the orthonormal vectors

$\vec{a}_1$ radial to $\{a^*\} \times \underline{\underline{R}}^{n-k}$,

$\vec{a}_2, \ldots, \vec{a}_k$ parallel to $R^k \times \{0\}$,

$\vec{a}_{k+1}$ radial to $\underline{\underline{R}}^k \times \{0\}$, and

$\vec{a}_{k+2}, \ldots, \vec{a}_n$ parallel to $\{0\} \times \underline{\underline{R}}^{n-k}$.

Then one may calculate

$$
\begin{aligned}
|Df(a)(x)(a_i)| \le\ & \delta^{-1} && \text{if } 1 \le i \le k \text{ and } k \in E_1(a), \\
& 0 && \text{if } i = 1 \text{ and } x \in E_2(a), \\
& 2r/|T(x) - a^*| && \text{if } 2 \le i \le k \text{ and } x \in E_2(a), \\
& (4r^2 + \xi^2)^{1/2}/\xi && \text{if } i = k+1 \text{ and } x \in E_4(a), \\
& 1 && \text{otherwise.}
\end{aligned}
$$

Thus, recalling 2.8,

$$
\begin{aligned}
\|f(a)_{\#}V\|(\underline{\underline{R}}^n) &= \int |\Lambda_k Df(a)(x) \circ S|\, dV(x,S) \\
&\le \int_{E_1(a)} r^k \delta^{-k} |\Lambda_k DT \circ S| + \delta^{1-k} r^k \, dV(x,S) \\
&+ \int_{E_3(a)} \|S - T\| [2r/|T(x) - a^*|]^{k-1} dV(x,S) \\
&+ \int_{E_4(a)} (4r^2 + \xi^2)^{1/2} \xi^{-1} \|S - T\| [2r/|T(x) - a^*|]^{k-1} dV(x,S) \\
&+ \|V\| [\underline{\underline{R}}^n \sim E_1(a) \cup E_2(a)] \ .
\end{aligned}
$$

Using (10) and (11) and various simplifications gives

$$\|f(a)_{\#}V\|(\underline{\underline{R}}^n) \le (\nu - 1 + 2\eta)\underline{\underline{\alpha}}r^k$$
$$+ (2r + \xi)\xi^{-1} \int \|S - T\| [2r/|T(x) - a^*|]^{k-1} dV(x,S)$$
$$+ \|V\| [E(r,\rho_1) \sim E(r,\rho_1 - \xi)].$$

Integrating this over all a in A yields

$$(12) \qquad \int_A \|f(a)_{\#}V\|(R^n) d\mathscr{H}^k a$$
$$\le \mathscr{H}^k(A)(\nu - 1 + 2\eta)\underline{\underline{\alpha}}r^k$$
$$+ (2r + \xi)\xi^{-1} \int \|S - T\| \int [2r/|T(x) - a^*|]^{k-1} d\mathscr{H}^k a\, dV(x,S)$$
$$+ \mathscr{H}^k(A)\|V\| [E(r,\rho_1) \sim E(r,\rho_1 - \xi)].$$

Now, since $a^* = (1 + \delta/r)a$ and $A \subset \underline{\underline{B}}^k(0,r - \delta)$, we have for fixed x

$$(13) \qquad \int_A [2r/|T(x) - a^*|]^{k-1} d\mathscr{H}^k a$$
$$< \int_{|T(x)-a^*|<2r} [2r/|T(x) - a^*|]^{k-1} (1 + \delta/r)^{-k} d\mathscr{H} a^*$$
$$< k2^k \underline{\underline{\alpha}}r^k .$$

It follows from (8) that

(14) $$\|V\|[E(r,\rho_1) \sim E(r,\rho_1 - \xi)] < k(\nu - \zeta)\underline{\underline{\alpha}}r^{k-1}\xi R/\rho \ .$$

Plugging (5), (13), and (14) into (12) yields

$$\int_A \|f(a)_\# V\|(\underline{\underline{R}}^n)\,d\mathscr{H}^k a$$

$$\leq \mathscr{H}^k(A)(\nu - 1 + 2\eta)\underline{\underline{\alpha}}r^k + (2r + \xi)\xi^{-1}\gamma\underline{\underline{\alpha}}r^k k 2^k \underline{\underline{\alpha}}r^k$$

$$+ \mathscr{H}^k(A)k(\nu - \zeta)\underline{\underline{\alpha}}r^{k-1}\xi R/\rho \ .$$

Therefore, using (9), we conclude that there is an $a \in A$ such that

$$\|f(a)_\# V\|(\underline{\underline{R}}^n) \leq [\nu - 1 + 2\eta + (1 + \eta)\zeta^{-1}\nu(2r + \xi)\xi^{-1}\gamma k 2^k$$

$$+ k(\nu - \zeta)\xi R/\rho r]\underline{\underline{\alpha}}r^k \ .$$

Recalling that η and ξ were arbitrary and $R/\rho < \mu^{-1}$, we may choose η, ξ/r, and γ depending only on ν, μ, k, and ζ so that $(1 - \zeta)/2 > \gamma$ and

(15) $$\|f(a)_\# V\|\underline{\underline{R}}^n \leq \|V\|\underline{\underline{R}}^n - 2^{-1}(1 - \gamma)\underline{\underline{\alpha}}r^k \ .$$

By 4.10 we shall have $f(a) \in \underline{\underline{E}}(V,\sigma,w)$ if

$$\|f(a)_\# V\|\underline{\underline{R}}^n \leq \exp[-w \operatorname{diam} E(r,\rho_1)]\|V\|\underline{\underline{R}}^n \ .$$

This is implied by

$$2^{-1}(1 - \zeta)(\nu - \zeta)^{-1} > 1 - \exp[-4w\sigma],$$

which is implied by hypothesis (1). Since $(1 - \zeta)/2 > \gamma$, (15) contradicts hypothesis (6). □

4.21. Larger radii.

The next two lemmas handle the case where the holes are large enough to be detected by the smoothed mean curvature. This first lemma is a slight modification of [AW1 6.1], having r_0 as the lower bound of radii instead of 0.

Lemma: Suppose

(1) $\nu \in \mathbb{N}$, $0 < \xi < 1$, $1 < M < \infty$, $0 < r_0 < R < \infty$, $T \in \mathbf{G}(n,k)$, and $V \in \mathbf{V}_k(\mathbb{R}^n)$;

(2) Y is a subset of $T^{\perp}$ with no more than $\nu + 1$ elements;

(3) $(M + 1)\operatorname{diam} Y = R$;

(4) $r_0 < (3\nu)^{-1}\operatorname{diam} Y$;

(5) $R\|\delta V\|\mathbf{B}(y,r) \le \xi\|V\|\mathbf{B}(y,r)$ whenever $y \in Y$ and $r_0 < r < R$; and

(6) $\int_{\mathbf{B}(y,r)} \|S - T\| dV(x,S) \le \xi\|V\|\mathbf{B}(y,r)$ whenever $y \in Y$ and $r_0 < r < R$.

Then there are $V_1, V_2 \in \mathbf{V}_k(\mathbf{R}^n)$ *and a partition of* Y *into subsets* Y_0, Y_1, Y_2 *such that*

(7) $V \geq V_1 + V_2$;

(8) *Neither* Y_1 *nor* Y_2 *has more than* ν *elements;*

(9) $(M \text{ diam } Y)\|\delta V_j\|\mathbf{B}(y,r) \leq 2M(\nu + 1)(3\nu M)^{k+1}(\exp \xi)\xi\|V\|\mathbf{B}(y,r)$

whenever $j = 1,2,\ y \in Y_j$ *and* $r_0 < r < M \text{ diam } Y$;

(10) $$\int_{\mathbf{B}(y,r)} \|S - T\| dV(x,S) \leq M(3\nu M)^k(\exp \xi)\xi\|V_j\|\mathbf{B}(y,r)$$

whenever $j = 1,2,\ y \in Y_j$ *and* $r_0 < r < M \text{ diam } Y$;

(11) $V_j \geq V \llcorner \{x \in \mathbf{R}^n : \text{dist}(T^{\perp}(x), Y_i) \leq r_0\}$ *whenever* $j = 1,2$; *and*

(12) $$[(1 + 1/M)^k + (\nu + 1)/M](\exp \xi)\,\frac{\|V\|\{x:\text{dist}(x,y) \leq R\}}{\alpha R^k}$$

$$\geq \sum \{\|V\|\mathbf{B}(y,r_0)/\alpha(k)r_0^k : y \in Y_0\}$$

$$+ \frac{\|V_1\|\{x:\text{dist}(x,Y_1) \leq M \text{ diam } Y\}}{\alpha(M \text{ diam } Y)^k}$$

$$+ \frac{\|V_2\|\{x:\text{dist}(z,Y_2) \leq M \text{ diam } Y\}}{\alpha(M \text{ diam } Y)^k} \quad . \ \square$$

4.22 Density ratios.

This lemma corresponds to [AW1 6.2]. It shows that a nearly flat varifold passing through several vertically separated points must have either several layers or a high rate of mass loss.

Lemma: Corresponding to each $1 < \lambda < 2$ and $\nu \in \underline{\underline{N}}$, there is $\gamma > 0$ with the following property: Suppose

(1) $V \in \underline{\underline{\Omega}}$, $T \in \underline{\underline{G}}(n,k)$, $Y \subset T^{\perp}$, Y has no more than ν elements, $\Theta^k(\|V\|,y) \in \underline{\underline{N}}$ for each $y \in Y$, $0 < \sigma < R < \infty$, $\operatorname{diam} Y < \gamma R$, $\varepsilon < \gamma^2\sigma$, $\omega > 0$ and $1/4\nu > 1 - \exp[-4\omega\sigma]$;

(2) $R\|\Phi_\varepsilon * \delta V\|\underline{\underline{B}}(y,r) \le \gamma\|\Phi_\varepsilon * V\|\underline{\underline{B}}(y,r)$ and

(3) $$\int_{\underline{\underline{B}}(y,r)} \|S - T\| d\Phi_\varepsilon * V(x,S) \le \gamma\|\Phi_\varepsilon * V\|\underline{\underline{B}}(y,r)$$

whenever $y \in Y$ and $\gamma\sigma < r < R$;

(4) $$\int_{\{(x,S):|T(x)|\le r,\operatorname{dist}(T^{\perp}(x),Y)<\sigma\}} \|S - T\| dV(x,S) < \gamma\underline{\underline{\alpha}}r^k$$

and

(5) $\Delta_{\sigma,\omega}\|V\|\{x:|T(x)| \le r, \operatorname{dist}(T^{\perp}(x),Y) < \sigma\} > -\gamma\underline{\underline{\alpha}}r^k$

for $0 < r < \sigma$.

Then

(6) $\lambda\|\Phi_\varepsilon * V\|\{x:\operatorname{dist}(x,Y) \le R\} \ge \underline{\underline{\alpha}}R^k \sum \{\Theta^k(\|V\|,y):y \in Y\}$.

Proof: It follows from repeated application of 4.21 and 4.17 to $\Phi_\varepsilon * V$ that there is a $\gamma_1 > 0$ such that if $0 < \gamma < \gamma_1$, $r_0 = \gamma\sigma$, and (1), (2) and (3) are satisfied, then there is a partition $Y_0, Y_1, Y_2, \ldots, Y_j$ of Y such that

(7) $\operatorname{diam} Y_i < \sigma$ for $i = 1,2,\ldots$ and

(8) $$\lambda^{1/4} R^{-k} \|\Phi_\varepsilon * V\| \{x : \operatorname{dist}(x,Y) \le R\}$$

$$\ge \sum \{ r_0^{-k} \|\Phi_\varepsilon * V\| \underline{\underline{B}}(Y, r_0) : y \in Y_0 \}$$

$$+ \sum_{i=1}^{j} \sigma^{-k} \|\Phi_\varepsilon * V\| \{x : \operatorname{dist}(T^\perp(x), Y_i) < r_0, |T(x)| < \sigma\}.$$

From the definition of $\Phi_\varepsilon * V$ and geometry, it follows that there exists $\gamma_2 > 0$ depending only on λ and ν such that if $\gamma < \gamma_2$ and $\varepsilon < \gamma^2\sigma$ then

(9) $$\lambda^{1/4} \|\Phi_\varepsilon * V\| \underline{\underline{B}}(y, \gamma\sigma)$$

$$\ge \|V\| \{x : |T(x)| < \gamma\sigma\lambda^{-1/4k}, |T^\perp(x) - y| < \gamma\sigma(1 - \lambda^{-1/4k})\}$$

for $y \in Y_0$, and

(10) $$\lambda^{1/4} \|\Phi_\varepsilon * V\| \{x : \operatorname{dist}(T^\perp(x), Y_i) < \gamma\sigma, |T(x)| \le \sigma\}$$

$$\ge \|V\| \{x : \operatorname{dist}(T^\perp(x), Y_i) < \gamma\sigma\lambda^{-1/4k}, |T(x)| \le \sigma\lambda^{-1/4k}\}$$

for $i = 1,2,\ldots,j$.

It follows from (7) and 4.20 that there is a $\gamma_3 > 0$ depending on λ and ν such that if $\gamma < \gamma_3$ and (4) and (5) are satisfied, then

(11) $$\lambda^{1/4}\|V\|\{x:|T(x)| < \gamma\sigma\lambda^{-1/4k}, |T^{\perp}(x) - y| < \gamma\sigma(1-\lambda^{-1/4k})\}$$

$$\geq \underline{\underline{\alpha}}\gamma^k\sigma^k\lambda^{-1/4}\Theta^k(\|V\|,y)$$

for $y \in Y_0$ and

(12) $$\lambda^{1/4}\|V\|\{x:\text{dist}(T^{\perp}(x),Y_i) < \gamma\sigma\lambda^{-1/4k}, |T(x)| \leq \sigma\lambda^{-1/4k}\}$$

$$\geq \sum \{\underline{\underline{\alpha}}\sigma^k\lambda^{-1/4}\Theta^k(\|V\|,y):y \in Y_i\}$$

for $i = 1,2,\ldots,j$.

Letting $\gamma = \min\{\gamma_1, \gamma_2, \gamma_3\}$ and combining (8), (9), (10), (11), and (12) gives the desired result. □

4.23. Integral density ratios.

This lemma is analogous to [AW1 6.3]. It shows that a nearly flat integral varifold must have a nearly integral number of layers all over.

Lemma: Suppose $V_1, V_2, \ldots \in \underline{\underline{I\Omega}}$, $0 < d < \infty$, $T \in \underline{\underline{G}}(n,k)$, $\sigma_i, \varepsilon_i, \omega_i > 0$ for $i = 1,2,3\ldots,$

(1) $$\lim_{i\to\infty} V_i = \lim_{i\to\infty} \Phi_{\varepsilon_i} * V_i = d\underline{\underline{v}}(T),$$

(2) $$\lim_{i \to \infty} \varepsilon_i/\sigma_i = \lim_{i \to \infty} \omega_i\sigma_i = 0 ,$$

and for some neighborhood W *of* 0

(3) $$\lim_{i \to \infty} \|\delta\Phi_{\varepsilon_i} * V_i\| W = 0 , \quad \textit{and}$$

(4) $$\lim_{i \to \infty} \Delta_{\sigma_i,\omega_i} \|V_i\| W = 0 .$$

Then d *is a nonnegative integer.*

Proof: Suppose ν is the smallest positive integer greater than d. Choose $1 < \lambda < \infty$ such that $\lambda^{k+2} d < \nu$. Let γ be as in 4.22. Choose $0 < R < \infty$ such that $\underline{\underline{B}}(0,(\lambda^2 + 4\gamma^2)R) \subset W$.

For each $i = 1,2,...$ let A_i be the set of those $x \in \underline{\underline{B}}(0,(\lambda - 1)R)$ such that $2|T^{\perp}(x)| < \gamma R$ and $\Theta^k(\|V_i\|,x)$ is a positive integer. Let B_i be the set of those $x \in A_i$ such that

$$R\|\delta\Phi_{\varepsilon_i} * V_i\| \underline{\underline{B}}(x,r) \le \gamma \|\Phi_{\varepsilon_i} * V_i\| \underline{\underline{B}}(x,r)$$

and

$$\int_{\underline{\underline{B}}(x,r)} \|S - T\| d\Phi_{\varepsilon_i} * V(y,S) < \gamma \|\Phi_{\varepsilon_i} * V_i\| B(x,r)$$

whenever $\sigma < r < R$. From the properties of convolution,

$$\|V_i\|(A_i - B_i) < (1 + \varepsilon_i/\sigma_i) \|\Phi_{\varepsilon_i} * V_i\| \{x : \operatorname{dist}(x, A_i - B_i) \le \sigma_i\}.$$

By the Besicovitch covering theorem 2.2,

$$\|\Phi_{\varepsilon_i} * V_i\|\{x : \operatorname{dist}(x, A_i - B_i) \le \sigma_i\}$$

$$\le \gamma^{-1}\underline{\underline{B}}(n)\,[R\|\delta\Phi_{\varepsilon_i} * V_i\|\underline{\underline{B}}(0,\lambda R) + \int_{\underline{\underline{B}}(0,\lambda R)} \|S - T\| d\Phi_{\varepsilon_i} * V_i(x,S)].$$

By hypotheses (1) and (3)

$$\lim_{i \to \infty} \int_{\underline{\underline{B}}(0,\lambda R)} \|S - T\| d\Phi_{\varepsilon_i} * V_i(x,S) = 0,$$

$$\lim_{i \to \infty} \|V_i\|[\underline{\underline{B}}(0,(\lambda - 1)R) \sim A_i] = 0 \text{ , and}$$

$$\lim_{i \to \infty} \|\delta\Phi_{\varepsilon_i} * V_i\|\underline{\underline{B}}(0, R) = 0.$$

Hence

$$\lim_{i \to \infty} \|V_i\|[\underline{\underline{B}}(0,(\lambda - 1)R) \sim B_i] = 0 \text{ ,}$$

and so

(5) $$\lim_{i \to \infty} V_i \llcorner B_i = d\underline{\underline{v}}[T \cap \underline{\underline{B}}(0,(\lambda - 1)R)].$$

For each $i = 1,2,\ldots$ let C_i be the set of $a \in T \cap \underline{\underline{B}}(0,(\lambda - 1)R)$ such that

$$\Delta_{\sigma_i,\omega_i}\|V_i\|\{x : |T(x - a)| < r,\ |T^{\perp}(x - a)| < 2\gamma R\} > -\gamma\underline{\underline{\alpha}}r^k$$

and

$$\int_{\{(x,S):|T(x-a)|<r,|T^{\perp}(x-a)|<2\gamma R\}} \|S - T\| dV(x,S) < \gamma \underline{\underline{\alpha}} r^k$$

whenever $0 < r < R$.

By the Besicovitch covering theorem,

$$\mathscr{H}^k[T \cap \underline{\underline{B}}(0,(\lambda - 1)R) \sim C_i]$$

$$\leq \gamma^{-1}\underline{\underline{B}}(k)[-\Delta_{\sigma_i,\omega_i}\|V_i\|B(0,(\lambda^2 + 4\gamma^2)^{1/2}R)$$

$$+ \int_{\underline{\underline{B}}(0,(\lambda^2+4\gamma^2)^{1/2}R)} \|S - T\| dV_i(x,S)].$$

By hypothesis,

$$\lim_{i \to \infty} \Delta_{\sigma_i,\omega_i}\|V_i\|\underline{\underline{B}}(0,(\lambda^2 + 4\gamma^2)^{1/2}R) = 0 \quad \text{and}$$

$$\lim_{i \to \infty} \int_{\underline{\underline{B}}(0,(\lambda^2+4\gamma^2)^{1/2}R)} \|S - T\| dV_i(x,S) = 0 ,$$

so, recalling (5),

$$\lim_{i \to \infty} V_i \llcorner B_i \cap T^{-1}[C_i] = d\underline{\underline{v}}[T \cap B(0,(\lambda - 1)R)],$$

which in turn implies that

(6) $$\lim_{i \to \infty} T_{\#}(V_i \llcorner B_i \cap T^{-1}[C_i] = d\underline{\underline{v}}[T \cap \underline{\underline{B}}(0,(\lambda - 1)R].$$

For each $z \in T$, let $Y_i(z) = A_i \cap T^{-1}[\{z\} \cap C_i]$. Inasmuch as

$$\lim_{i \to \infty} \|\Phi_{\varepsilon_i} * V_i\| \underline{\underline{B}}(0,\lambda R) = d\underline{\underline{\alpha}}(\lambda R)^k ,$$

we see that for large i

$$\|\Phi_{\varepsilon_i} * V_i\| \{x : \mathrm{dist}(x, Y_i(z)) < R\} < \lambda^{k+1} d\underline{\underline{\alpha}} R^k$$

for all $z \in T$. By choice of λ and 4.22, we see that for large i

$$\sum \{\Theta^k(\|V_i\|, y) : y \in Y\} \le \lambda^{k+2} d < \nu$$

whenever $z \in T$ and Y is a subset of $Y_i(z)$ consisting of no more than ν elements. Therefore, if i is sufficiently large,

$$\sum \{\Theta^k(\|V_i\|, y) : y \in Y_i(z)\} \le \nu - 1$$

for all $z \in T$. The definition of mapping varifolds and the properties of C_i imply

$$\|T_{\#}(V_i \llcorner B_i \cap T^{-1}[C_i]\| \underline{\underline{R}}^n = \int_T \sum \{\Theta^k(\|V_i\|, y) : y \in Y_i(z)\} d\mathscr{H}^k z$$

$$\le (\nu - 1)\mathscr{H}^k[C_i] \le (\nu - 1)\underline{\underline{\alpha}}((\lambda - 1)R)^k.$$

This combined with (6) implies $d = \nu - 1$. □

4.24. Integrality

We conclude the first part proof of integrality with this adaptation of [AW1 6.4].

Theorem: Suppose $0 < B < \infty$, $V_1, V_2, \ldots \in \underline{\underline{I}}\Omega$,

(1) $$\Omega \lim_{m \to \infty} V_m = V \in \underline{\underline{\Omega V}},$$

(2) $$\|V_m\|(\Omega) < B, \quad \text{and}$$

(3) $$\langle \Omega, |\Phi_{\varepsilon(m)} * \delta V_m|^2 / \Phi_{\varepsilon(m)} * \|V_m\| \rangle - \Delta_{\sigma(m),m} \|V_m\|(\Omega)/\Delta t(m) < B$$

for all $m \in \underline{\underline{N}}$. Then V is integral.

Proof: From 4.19 we know $V \in \underline{\underline{RV}}_k(\underline{\underline{R}}^n)$. For each pair of positive integers m and q, let $A_{m,q}$ be the set consisting of all $x \in \underline{\underline{R}}^n$ such that

(4) $$\|\delta\Phi_{\varepsilon(m)} * V_m\| \underline{\underline{B}}(x,r) < q \|\Phi_{\varepsilon(m)} * V\| \underline{\underline{B}}(x,r)$$

whenever $\sigma < r < 1$, and

(5) $$\Delta_{\sigma(m),m} \|V_m\| \underline{\underline{B}}(x,r) > -q \Delta t(m) \|V_m\| \underline{\underline{B}}(x,r)$$

whenever $0 < r < 1$. By using Schwarz' inequality, (2) and (3) yield

(6) $$\|\delta\Phi_{\varepsilon(m)} * V_m\|(\Omega) \le [\langle \Omega, |\Phi_{\varepsilon(m)} * V_m|^2 / \Phi_{\varepsilon(m)} * V_m \rangle \|\Phi_{\varepsilon(m)} * V_m\|(\Omega)]^{1/2}$$
$$\le B .$$

The Besicovitch covering theorem, (6), and the properties of Ω imply

$$\|V_m\|(\Omega \llcorner \mathbf{R}^n \sim A_{m,q}) < 8q^{-1}\mathbf{B}(n)B. \tag{7}$$

Let A consist of all $x \in \mathbf{R}^n$ such that for some $q \in \mathbf{N}$ there are $x_m \in A_{m,q}$ for infinitely many m with $x = \lim_{m \to \infty} x_m$. Then (7) implies

$$\|V\|(\Omega \llcorner \mathbf{R}^n \sim A) = 0. \tag{8}$$

Let A^* consist of all $x \in A$ such that

$$0 < \Theta^k(\|V\|, x) < \infty ,$$

$$\operatorname{Tan}^k(\|V\|, x) \in \mathbf{G}(n,k), \quad \text{and}$$

$$\lim_{r \to \infty} \boldsymbol{\mu}(r) \circ \tau(-x)_\# V = \Theta^k(\|V\|, x)\,\mathbf{v}[\operatorname{Tan}^k(\|V\|, x)] .$$

Since $V \in \mathbf{RV}_k(\mathbf{R}^n)$ it follows from (8) and [AW1 3.5(1)] that

$$\|V\|(\Omega \llcorner R^n \sim A^*) = 0. \tag{9}$$

Let $a \in A^*$, and let $q \in \mathbf{N}$ and $a_1, a_2, \ldots$ be such that $\lim_{m \to \infty} a_m = a$ and $a_m \in A_{m,q}$. For each positive integer j choose $m(j)$ such that $|a - a_{m(j)}| < j^{-1}$ and

$$\begin{aligned} \lim_{j \to \infty} (\boldsymbol{\mu}(j) \circ \boldsymbol{\tau}(-x))_\# V_{m(j)} &= \lim_{j \to \infty} (\boldsymbol{\mu}(j) \circ \boldsymbol{\tau}(-x))_\# \Phi_{\varepsilon(m(j))} {}^* V_{m(j)} \\ &= \Theta^k(\|V\|, a)\,\mathbf{v}[\operatorname{Tan}^k(\|V\|, a)] . \end{aligned} \tag{10}$$

With a view to applying 4.23 to $\underline{\underline{\mu}}(j) \circ \underline{\underline{\tau}}(-a)_{\#} V_{m(j)}$, we calculate

$$\limsup_{j \to \infty} \|\delta\Phi_{j\varepsilon(m(j))} * \underline{\underline{\mu}}(j) \circ \underline{\underline{\tau}}(-a)_{\#} V_{m(j)}\| \underline{\underline{U}}(0,1)$$

$$= \limsup_{j \to \infty} \|\delta(\underline{\underline{\mu}}(j) \circ \underline{\underline{\tau}}(-a))_{\#}(\Phi_{\varepsilon(m(j))} * V_{m(j)})\| \underline{\underline{U}}(0,1)$$

$$= \limsup_{j \to \infty} j^{k-1} \|\delta\Phi_{\varepsilon(m(j))} * V_{m(j)}\| \underline{\underline{U}}(a, j^{-1})$$

$$\le \limsup_{j \to \infty} j^{k-1} \|\delta\Phi_{\varepsilon(m(j))} * V_{m(j)}\| \underline{\underline{B}}(a_{m(j)}, 2j^{-1})$$

$$\le \limsup_{j \to \infty} j^{k-1} q \|\Phi_{\varepsilon(m(j))} * V_{m(j)}\| \underline{\underline{B}}(a_{m(j)}, 2j^{-1})$$

$$\le \limsup_{j \to \infty} \quad j^{k-1} q 2^k j^{-k} \underline{\underline{\alpha}} \Theta^k(\|V\|, a) = 0 \quad ,$$

where we used (5) and (10) at the end. Also, by (6) and (10),

$$\limsup_{j \to \infty} -\Delta_{j\sigma(m(j)), m(j)/j} \|(\underline{\underline{\mu}}(j) \circ \underline{\underline{\tau}}(-a))_{\#} V_{m(j)}\| \underline{\underline{U}}(0,1)$$

$$= \limsup_{j \to \infty} -j^k \Delta_{\sigma(m(j)), m(j)} \|V_{m(j)}\| \underline{\underline{U}}(a, j^{-1})$$

$$\le \limsup_{j \to \infty} -j^k \Delta_{\sigma(m(j)), m(j)} \|V_{m(j)}\| \underline{\underline{B}}(a_{m(j)}, 2j^{-1})$$

$$\le \limsup_{j \to \infty} j^k q \|V_{m(j)}\| \underline{\underline{B}}(a_{m(j)}, 2j^{-1}) \Delta t(m(j))$$

$$\le \limsup_{j \to \infty} j^k q 2^k j^{-k} \Theta^k(\|V\|, a) \Delta t(m(j)) = 0.$$

Applying 4.23 to $(\underline{\underline{\mu}}(j)\circ\underline{\underline{\tau}}(-a))_{\#}V_{m(j)}$ with $\sigma_j = j\sigma(m(j))$, $\varepsilon_j = j\varepsilon(m(j))$, and $\omega_j = m(j)/j$, we conclude that $\Theta^k(\|V\|,a)$ must be a positive integer. Since this is true for $\|V\|$ almost all $a \in \underline{\underline{R}}^n$ by (9), we have $V \in \underline{\underline{IV}}_k(\underline{\underline{R}}^n)$. □

4.25 Times of good behavior.

As noted earlier, most of our estimates are in terms of rate of mass loss. Therefore we are very interested in times where the rate of mass loss is small.

Suppose $V_0 \in \underline{\underline{\Omega}}$, and let $V_{m,t}$ and $\|V_t\|$ be as defined in 4.13 and 4.14. For every pair of positive integers m and q define

(1) $$\underline{\underline{P}}(q,m) = \{t \in \underline{\underline{Q}}_m : \langle \Omega, |\Phi_{\varepsilon(m)} * \delta V_{m,t}|^2/\Phi_{\varepsilon(m)} * \|V_{m,t}\|\rangle$$

$$- \Delta_{\sigma(m),m}\|V_{m,t}\|(\Omega)/\Delta t(m) < q\},$$

(2) $$\underline{\underline{PP}}(q,m) = \cup\{[t,t+\Delta t(m)) : t \in \underline{\underline{P}}(q,m)\},$$ and

(3) $\underline{\underline{PP}}(q) = \{t \in \underline{\underline{R}}^+$: for all $\eta > 0$ and $M \in N$ there exist $m \in N$ and $s \in \underline{\underline{P}}(q,m)$ such that $m > M$ and $|t - s| < \eta\}$.

Proposition: Suppose $V_0 \in \underline{\underline{\Omega}}$. Then:

(a) If $m, q \in \underline{\underline{N}}$, $s, t \in \underline{\underline{Q}}_m$, $s < t$, and

(4) $$\|V_{m,s}\|(\Omega) < (q/16)\exp(s - t),$$

then

(5) $$\mathcal{L}^1([s,t] \sim \underline{\underline{PP}}(m,q))$$

$$\leq (2/q)(1 - m^{-4})^{-1}[\|V^*_{m,s}\|(\Omega) - \|V^*_{m,t}\|(\Omega)$$

$$+ 2(1 + m^{-4})\|V^*_{m,s}\|(\Omega)(t - s)\exp 2(t - s)].$$

(b) $$\mathcal{L}^1(\underline{\underline{R}}^+ \sim \bigcup_q \underline{\underline{PP}}(q)) = 0.$$

Proof: If $r \in [s,t] \cap \underline{\underline{Q}}_m$, then by 4.13 and 4.12(i)

(6) $$[\|V^*_{m,r+\ t(m)}\|(\Omega) - \|V^*_{m,r}\|(\Omega)]/\Delta t(m)$$

$$\leq (-1 + m^{-4})\langle \Omega, |\Phi_{\varepsilon(m)} * \delta V^*_{m,r}|^2/\Phi_{\varepsilon(m)} * \|V^*_{m,r}\|\rangle$$

$$+ (1 + m^{-4})\langle \Omega, |\Phi_{\varepsilon(m)} * \delta V^*_{m,r}|^2/\Phi_{\varepsilon(m)} * \|V^*_{m,r}\|\rangle^{1/2}\|V^*_{m,r}\|(\Omega)^{1/2}$$

$$+ m^{-4}\|V^*_{m,r}\|(\Omega) + (1 - 2^{-m})\Delta_{\sigma(m),m}\|V_{m,r}\|(\Omega)/\Delta t(m).$$

By 4.12(ii) and hypothesis (4) we have $\|V^*_{m,r}\|(\Omega) < q/16$. Thus whenever

(7) $$\langle \Omega, |\Phi_{\varepsilon(m)} * V_{m,r}|^2/\Phi_{\varepsilon(m)} * \|V_{m,r}\|\rangle - \Delta_{\sigma(m),m}\|V_{m,r}\|(\Omega)/\Delta t(m) \geq q$$

we can infer from (6) and 4.11(iii) that

(8) $$[\|V^*_{m,r+\Delta t(m)}\|(\Omega) - \|V^*_{m,r}\|(\Omega)/\Delta t(m) \leq -(1 - m^{-4})q/2.$$

Since by 4.12(ii) we have for all $r \in (s,t) \cap \underline{\underline{Q}}_m$

$$(9) \quad [\|V^*_{m,r+\Delta t(m)}\|(\Omega) - \|V^*_{m,r}\|(\Omega)]/\Delta t(m) \leq 2\|V^*_{m,r}\|(\Omega)$$

$$\leq \|V^*_{m,s}\|(\Omega) \exp 2(t - s) \ ,$$

we must have, using (8) for $r \notin \underline{\underline{PP}}(q,m)$ and (9) for $r \in \underline{\underline{PP}}(q,m)$,

$$\|V^*_{m,s}\|(\Omega) - (1 - m^{-4})(q/2)\mathscr{L}^1([s,t] \sim \underline{\underline{PP}}(q,m))$$

$$+ (t - s)(1 + m^4) 2\|V_{m,s}\|(\Omega) \exp 2(t - s)$$

$$\geq \|V^*_{m,t}\|(\Omega).$$

This implies

$$\mathscr{L}^1([s,t] \sim \underline{\underline{PP}}(q,m))$$

$$\leq 2q^{-1}(1 - m^{-4})^{-1}\{\|V^*_{m,s}\|(\Omega) - \|V^*_{m,t}\|(\Omega)$$

$$+ 2(t - s)(1 + m^4)\|V^*_{m,s}\|(\Omega) \exp 2(t - s)\},$$

which proves (a).

It follows from 4.12(ii) that for any $m \in \underline{\underline{N}}$ with $t \in \underline{\underline{Q}}_m$,

$$(10) \qquad \|V_{m,t}\|(\Omega) \leq \exp(2t)\|V_0\|(\Omega).$$

Thus for $q \in \underline{\underline{N}}$ such that $q > 16 \exp(2t)\|V_0\|(\Omega)$ we get from (5) that

$$\mathcal{L}^1([0,t] \sim \underline{\underline{PP}}(q,m)) \le 4q^{-1}[1 + 2te^{2t}]\|V_0\|(\Omega) .$$

Recalling the definition (3) of $\underline{\underline{PP}}(q)$, we see that

$$\mathcal{L}^1([0,t] \sim \underline{\underline{PP}}(q)) \le \liminf_{m \to \infty} \mathcal{L}^1([0,t] \sim \underline{\underline{PP}}(q,m))$$
$$< 4q^{-1}[1 + 2te^{2t}]\|V_0\|(\Omega) .$$

Hence $\mathcal{L}^1([0,t] \sim \bigcup_q \underline{\underline{PP}}(q)) = 0$ for all $t > 0$, which proves (b). □

4.26. Definition of V_t and basic properties.

We now make $\|V_t\|$ determine V_t whenever possible, which we shall show is almost always.

Definition: Suppose $V_0 \in \underline{\underline{\Omega}}$ and $\|V_t\|$ is as defined in 4.14(a). Let $T \in \underline{\underline{G}}(n,k)$ be arbitrary. For any $t > 0$ define $V_t \in \underline{\underline{\Omega V}}$ by

$$(1) \quad V_t(A) = \|V_t\|\{x:(x,\operatorname{Tan}^k(\|V\|_t,x)) \in A\}$$
$$+ \|V_t\|\{x:\operatorname{Tan}^k(\|V\|_t,x) \notin \underline{\underline{G}}(n,k) \text{ and } (x,T) \in A\} ,$$

whenever $A \subset \underline{\underline{G}}_k(\underline{\underline{R}}^n)$.

By [AW1 3.5(1)(a)], the second quantity on the right hand side of (1) is zero whenever V_t is rectifiable.

<u>Theorem</u>: (a) <u>If</u> $V_0 \in \underline{\underline{\Omega}}$, <u>then</u> V_t <u>is rectifiable for almost all</u> $t > 0$.

(b) <u>If</u> $V_0 \in \underline{\underline{I\Omega}}$, <u>then</u> V_t <u>is integral for almost all</u> $t > 0$.

<u>Proof</u>: Suppose $q \in \underline{\underline{N}}$, $t \in \underline{\underline{PP}}(q)$, and $\|V_t\|$ is continuous at t. Then by the definition 4.25(3) of $\underline{\underline{PP}}(q)$ there exist sequences m_i and t_i, $i = 1,2,\ldots$ such that $m_i \in \underline{\underline{N}}$, $\lim_{i \to \infty} m_i = \infty$, $t_i \in P(q,m)$, and $t = \lim_{i \to \infty} t_i$. By 4.14(g) we have

$$(2) \qquad \|V_t\| = \Omega \lim_{i \to \infty} \|V_{m_i,t_i}\| \ .$$

Since

$$\{ V \in \underline{\underline{\Omega V}}: \|V\|(\Omega) \leq \|V_t\|(\Omega) + 1\}$$

is compact by 4.1, any sequence $\langle V_{m_i,t_i}\rangle_{i=1}^{\infty}$ will have a convergent subsequence, and the limit W of this subsequence will be rectifiable by 4.25(1) and 4.19. Being rectifiable, by [AW1 3.5(1)] W is determined by $\|W\|$, which is $\|V_t\|$ by (2). Hence all subsequences have the same limit W, so $\lim_{i \to \infty} V_{m_i,t_i} = W$. Since $\|W\| = \|V_t\|$ and W is rectifiable, W is the same as the V_t defined by (1). By 4.25(b), almost every $t > 0$ is in some $\underline{\underline{PP}}(q)$, and by 4.14(f) $\|V_t\|$ is continuous at almost all t, so V_t is rectifiable for almost all $t > 0$.

If $V_0 \in \underline{\underline{I\Omega}}$, then the same argument as for (a) with 4.24 replacing 4.19 shows that V_t is integral for almost all $t > 0$. □

4.27 Motion on non-compact test functions

In this section we establish the inequality used to define motion by mean curvature on our sets $\mathscr{A}_i$ of test functions. Compact support test functions are the subject of 4.30.

Proposition: If $V_0 \in \underline{\underline{\Omega}}$, V_t is as defined in 4.26, $i \in \underline{\underline{N}}$, and $\phi \in \mathscr{A}_i$, then for almost all $t > 0$

(1) $$\bar{D}\|V_t\|(\phi) \le \delta(V_t,\phi)(\underline{\underline{h}}(V_t,\cdot)).$$

Proof: Suppose $q \in \underline{\underline{N}}$, $t \in \underline{\underline{PP}}(q)$, $d\|V_t\|(\omega)/dt > -q$, and $\|V_t\|(\Omega) < q/16$.

Let $\eta > 0$. Choose $j \in N$ so that

$$(1 - 50q/j)(\delta(V_t,\phi)(\underline{\underline{h}}(V_t,\cdot)) + 2\eta/3) + (50q/j)3i^2\|V_t\|(\phi)$$

$$\le \delta(V_t,\phi)(\underline{\underline{h}}(V_t,\cdot)) + \eta$$

Suppose $t_m \in \underline{\underline{P}}(m,j)$ and $\lim_{m \to \infty} t_m = t$. As was shown in the proof of 4.26,

(2) $$\Omega \lim_{m \to \infty} V_{m,t_m} = V_t$$

and V_t is rectifiable.

From 4.18 and 4.7(i) we have

(3) $$\int |\underline{\underline{h}}(V_t,x)|^2\phi(x)d\|V_t\|x$$

$$\le \liminf_{m\to\infty} \langle \phi, |\Phi_{\varepsilon(m)} * V_{m,t_m}|^2/\Phi_{\varepsilon(m)} * \|V_{m,t_m}\|\rangle \le j ,$$

$$\lim_{m\to\infty} |\delta V_{m,t_m}(\phi h_{\varepsilon(m)}(V_{m,t_m}))$$

$$+ \langle \phi, |\Phi_{\varepsilon(m)} * V_{m,t_m}|^2/\Phi_{\varepsilon(m)} * \|V_{m,t_m}\|\rangle| = 0,$$

and hence

(4) $$-\int |\underline{\underline{h}}(V_t,x)|^2\phi(x)d\|V_t\|x \ge \limsup_{m\to\infty} \delta V_{m,t_m}(\phi h_{\varepsilon(m)}(V_{m,t_m})).$$

Since $\int S^{\perp}(D\phi(\cdot))dV_t^{(\cdot)}S$ is a $\|V_t\|$ measurable vectorfield, there are $\tau \in \underline{\underline{N}}$ and $g \in \mathscr{B}_\tau$ such that

(5) $$\int |\int S^{\perp}(D\phi(x))dV_t^{(x)}S - g(x)|^2\phi(x)^{-1}d\|V_t\|x < \eta^2/16j \quad .$$

Since V_t is rectifiable, at $\|V_t\|$ almost all x there is a unique tangent plane, so

(6) $$\int |S^{\perp}(D\phi(x)) - g(x)|^2\phi(x)^{-1}dV_t(x,S) < \eta^2/16j \quad .$$

It follows from (2) that

$$(7)\qquad \lim_{m\to\infty} \int |S^{\perp}(D\phi(x)) - g(x)|^2 \phi(x)^{-1} dV_{m,t_m}(x,S) < \eta^2/16j.$$

It follows from (2), (3) and 4.8 that

$$(8)\qquad \lim_{m\to\infty} \int h_{\varepsilon(m)}(V_{m,t_m})(x)\cdot g(x)\, d\|V_{m,t_m}\|x$$

$$= \int \underline{\underline{h}}(V_t,x)\cdot g(x)\, d\|V_t\|x.$$

We may infer from (3), (5), (6), (7), and (8) that

$$(9)\qquad \int S^{\perp}(D\phi(x))\cdot \underline{\underline{h}}(V_t,x)\, dV_t(x,S) + \eta/2$$

$$\ge \limsup_{m\to\infty} \int S^{\perp}(D\phi(x))\cdot h_{\varepsilon(m)}(V_{m,t_m})(x)\, dV_{m,t_m}(x,S).$$

Together, (3) and (9) say that

$$(10)\qquad \delta(V_t,\phi)(\underline{\underline{h}}(V_t,\cdot)) + \eta/2 \ge \limsup_{m\to\infty} \delta(V_{m,t_m},\phi)(h_{\varepsilon(m)}(V_{m,t_m})).$$

It follows from the preceding argument, 4.9, 4.11(i), and 4.11(ii) that there are $M_1 \in \underline{\underline{N}}$ and $R_1 > 0$ such that if $m > M_1$, $|t - r| < R_1$ and $r \in \underline{\underline{P}}(m,j)$ then

$$(11)\qquad [\|V^*_{m,r+\Delta t(m)}\|(\phi) - \|V^*_{m,r}\|(\phi)]/\Delta t(m)$$

$$< \delta(V_t,\phi)(\underline{\underline{h}}(V_t,\cdot)) + 2\eta/3 ,$$

and if $r \notin P(m,j)$ then by 4.12(ii)

(12) $$[\|V^*_{m,r+\Delta t(m)}\|(\phi) - \|V^*_{m,r}\|(\phi)]/\Delta t(m) < 3i^2\|V_t\|(\phi).$$

Since $d\|V_t\|(\Omega)/dt > -q$, there are $M_2 \in \underline{\underline{N}}$ and $R_2 > 0$ such that if $m > M_2$, $|r - t| < R_1$, and $r \in \underline{\underline{Q}}_m$, then

(13) $$\|V^*_{m,t}\|(\Omega) < q/16 , \quad \text{and}$$

(14) $$\|V^*_{m,r}\|(\Omega) - \|V^*_{m,t}\|(\Omega)| < q|t - r| .$$

Suppose $M = \max\{M_1, M_2\}$ and $R = \min\{R_1, R_2, \frac{1}{10}\}$, and suppose $m > M$, $t - R < r < s < t + R$, and $r, s \in \underline{\underline{Q}}_m$. By (4.25(a), (13), and (14),

(15) $$\begin{aligned}&\mathscr{L}^1([r,s) \sim \underline{\underline{PP}}(m,j))\\ &\le (2/j)(1 - m^{-4})^{-1}[q(s - r)\\ &\quad + 2(1+m^{-4})(q/16)(s-r)\exp 2(s-r)]\\ &< 50(s - r)q/j.\end{aligned}$$

Combining (11), (12), and (15) yields

$$\begin{aligned}&[\|V^*_{m,s}\|(\phi) - \|V^*_{m,r}\|]/(s - r)\\ &\le (1 - 50\, q/j)(\delta(V_t,\phi)(\underline{\underline{h}}(V_t,\cdot)) + 2\eta/3) + (50q/j)3i^2\|V_t\|(\phi)\\ &\le \delta(V_t,\phi)(\underline{\underline{h}}(V_t,\cdot)) + \eta.\end{aligned}$$

Hence

$$[\|V_s\|(\phi) - \|V_r\|(\phi)]/(s - r) \le \delta(V_t,\phi)(\underline{\underline{h}}(V_t,\cdot)) + \eta,$$

and so, since η was arbitrary,

$$\bar{D}\|V_t\|(\phi) \le \delta(V_t,\phi)(\underline{\underline{h}}(V_t,\cdot)). \qquad \square$$

4.28. <u>Upper semicontinuity of</u> $\delta(V,\psi)(\underline{\underline{h}}(V,\cdot))$ <u>in</u> V.

The final step in this chapter is to pass from test functions without compact support to those with compact support. But first we prove a semicontinuity result that will be needed for that final step.

<u>Lemma:</u> <u>If</u> $\psi \in \underline{\underline{C}}_0^2(\underline{\underline{R}}^n,\underline{\underline{R}}^+)$, $V_0, V_1, \ldots \in \underline{\underline{V}}_k(\underline{\underline{R}}^n)$, $V_0 \llcorner \psi \in \underline{\underline{RV}}_k(\underline{\underline{R}}^n)$, $\limsup_{i\to\infty} \|V_i\|\{x:\psi(x) > 0\} < \infty$ <u>and</u> $\lim_{i\to\infty} V_i \llcorner \psi = V_0 \llcorner \psi$, <u>then</u>

$$\limsup_{i\to\infty} \delta(V_i,\psi)(\underline{\underline{h}}(V_i,\cdot)) \le \delta(V_0,\psi)(\underline{\underline{h}}(V_0,\cdot)).$$

<u>Proof:</u> Suppose the conclusion were false. Then we could choose $\eta > 0$ and a subsequence of V_i (labeled the same) such that

$$\text{(1)} \qquad \lim_{i\to\infty} \delta(V_i,\psi)(\underline{\underline{h}}(V_i,\cdot)) > \delta(V_0,\psi)(\underline{\underline{h}}(V_0,\cdot)) + \eta.$$

It follows from (1), 3.4, and the weak continuity of δV in V that there is $B < \infty$ such that

(2) $$B \geq \limsup_{i \to \infty} \int |\underline{\underline{h}}(V_i,x)|^2 \psi(x)\,d\|V_i\|x$$

$$\geq \int |\underline{\underline{h}}(V_0,x)|^2 \psi(x)\,d\|V_0\|x.$$

Assume also that B satisfies

(3) $$\limsup_{i \to \infty} \|V_i\|\{x:\psi(x) > 0\} < B.$$

We cannot prove semicontinuity for ψ directly. Therefore we use the finiteness of $\int |\underline{\underline{h}}|^2 \psi d\|V_0\|$ to choose $\phi \in \underline{\underline{C}}_0^2\ (\underline{\underline{R}}^n,\underline{\underline{R}}^+)$ such that $\phi \leq \psi$, spt $\phi \subset \{x:\psi(x) > 0\}$,

(4) $$\delta(V_0,\ \psi - \phi)(\underline{\underline{h}}(V_0,\cdot)) > -\eta/4,$$

and

(5) $$\sup\{|D\psi(x) - D\phi(x)|^2/|\psi(x) - \phi(x)|:x \in \underline{\underline{R}}^n\} < \eta/4B.$$

Note that (3), (5) and 3.4 imply

(6) $$\limsup_{i \to \infty} \delta(V_i,\ \psi - \phi)(\underline{\underline{h}}(V_i,\cdot)) < \eta/4\ .$$

Now we study $\delta(V_i,\phi)(\underline{\underline{h}}(V_i,\cdot))$.

It again follows from the weak continuity of δV in V that

(7) $$\int |\underline{\underline{h}}(V_0,x)|^2 \phi(x)\,d\|V_0\|x \leq \liminf_{i \to \infty} \int |\underline{\underline{h}}(V_i,x)|^2 \phi(x)\,d\|V_i\|x.$$

Since $\int S^{\perp}(D\phi(\cdot))dV_0^{(\cdot)}S$ is a $\|V_0\|$ measurable vectorfield and $\operatorname{spt}\phi \subset \operatorname{spt}\psi$, there is $g \in \underline{\underline{C}}_0^1(\underline{\underline{R}}^n,\underline{\underline{R}}^n)$ such that $\operatorname{spt} g \subset \operatorname{spt}\phi$ and

$$\int \left| \int S^{\perp}(D\psi(x))dV_0^{(x)}S - g(x) \right|^2 \psi(x)^{-1} d\|V_0\|x < \eta^2/16B ,$$

where we take the integrand to be 0 when $\psi(x) = 0$. Since $V_0 \llcorner \psi$ is rectifiable, at $\|V_0\| \llcorner \psi$ almost all x there is a unique tangent plane, and so

$$\int \left| \int S^{\perp}(D\phi(x))dV_0^{(x)}S - g(x) \right|^2 \psi(x)^{-1} d\|V_0\|x$$

$$= \int |S^{\perp}(D\phi(x)) - g(x)|^2 \psi(x)^{-1} dV_0(x,S).$$

Now $|S^{\perp}(D\phi(x)) - g(x)|^2/\psi(x)$ is a continuous function on $\underline{\underline{G}}_k(\underline{\underline{R}}^n)$ with support in $\{x \in \underline{\underline{R}}^n : \psi(x) > 0\} \times \underline{\underline{G}}(n,k)$, so

$$(8) \qquad \lim_{i\to\infty} \int |S^{\perp}(D\phi(x)) - g(x)|^2 \psi(x)^{-1} dV_i(x,S)$$

$$= \int |S^{\perp}(D\phi(x)) - g(x)|^2 \psi(x)^{-1} dV_0(x,S)$$

$$< \eta^2/16B.$$

Since $\operatorname{spt} g \subset \{x \in \underline{\underline{R}}^n : \psi(x) > 0\}$ we have

$$(9) \qquad \int g(x)\cdot\underline{h}(V_0,x)d\|V_0\|x = \delta V_0(g) = \lim_{i\to\infty} \delta V_i(g)$$

$$= \lim_{i\to\infty} \int g(x)\cdot\underline{h}(V_i,x)d\|V_i\|x.$$

We may calculate, using Schwarz' inequality, (2) and (8),

$$(10)\qquad \int S^{\perp}(D\phi(x))\cdot\underline{\underline{h}}(V_0,x)\,dV_0(x,S)$$

$$= \int g(x)\cdot\underline{\underline{h}}(V_0,x)\,dV_0(x,S) + \int S^{\perp}(D\phi(x)) - g(x))\cdot\underline{\underline{h}}(V_0,x)\,dV_0(x,S)$$

$$\le \lim_{i\to\infty}\int g(x)\cdot\underline{\underline{h}}(V_i,x)\,dV_i(x,S) + [\int |S^{\perp}(D\phi(x)) - g(x)|^2\psi(x)^{-1}dV_0(x,S)]^{1/2} \cdot [\int |\underline{\underline{h}}(V_0,x)|^2\psi(x)\,d\|V_0\|x]^{1/2}$$

$$\le \limsup_{i\to\infty}\int S^{\perp}(D\phi(x))\cdot\underline{\underline{h}}(V_i,x)\,dV_i(x,S) + \limsup_{i\to\infty}\int (g(x) - S^{\perp}(D\phi(x)))\cdot\underline{\underline{h}}(V_i,x)\,dV_i(x,S) + (\eta^2/16B)^{1/2}B^{1/2}$$

$$\le \limsup_{i\to\infty}\int S^{\perp}(D\phi(x))\cdot\underline{\underline{h}}(V_i,x)\,dV_i(x,S) + \limsup_{i\to\infty}\int |S^{\perp}(D\phi(x)) - g(x)|^2\psi(x)^{-1}dV_i(x,S)]^{1/2} \cdot [\int |\underline{\underline{h}}(V_i,x)|^2\psi(x)\,dV_i(x,S)]^{1/2} + \eta/4$$

$$\le \limsup_{i\to\infty}\int S^{\perp}(D\phi(x))\cdot h(V_i,x)\,dV_i(x,S) + \eta/4 + \eta/4.$$

Combining (7) and (10) gives

$$(11) \qquad \delta(V_0,\phi)(\mathbf{h}(V_0,\cdot)) > \limsup_{i \to \infty} \delta(V_i,\phi)(\mathbf{h}(V_i,\cdot)) - \eta/2 \quad .$$

Finally, combining (4), (6), and (11) gives

$$\begin{aligned}
\delta(V_0,\psi)(\mathbf{h}(V_0,\cdot)) &= \delta(V_0,\psi - \phi)(\mathbf{h}(V_0,\cdot)) + \delta(V_0,\phi)(\mathbf{h}(V_0,\cdot)) \\
&> -\eta/4 + \limsup_{i \to \infty} \delta(V_i,\phi)(\mathbf{h}(V_i,\cdot)) - \eta/2 \\
&> -\eta/4 + \limsup_{i \to \infty} [\delta(V_i,\psi)(\mathbf{h}(V_i,\cdot)) \\
&\qquad - \delta(V_i,\psi - \phi)(\mathbf{h}(V_i,\cdot))] - \eta/2 \\
&> -\eta + \limsup_{i \to \infty} \delta(V_i,\psi)(\mathbf{h}(V_i,\cdot)),
\end{aligned}$$

which contradicts (1). □

4.29. *Motion on compact test functions.*

Proposition 4.27 applies to almost all times and to test functions in the sets $\mathscr{A}_i$. We now establish the key inequality for all times and for test functions with compact support.

Theorem: *If* $V_0 \in \mathbf{\Omega}$, V_t *is as defined in* 4.26, $\psi \in \mathbf{C}_0^2(\mathbf{R}^n,\mathbf{R}^+)$ *and* $t > 0$ *then*

(a) *if* $\bar{D}\|V_t\|(\psi) > -\infty$ *then* $V_t \llcorner \{(x,S): \psi(x) > 0\}$ *is rectifiable,*

(b) *if* $V_0 \in \mathbf{I\Omega}$ *and* $\bar{D}\|V_t\|(\psi) > -\infty$ *then* $V_t \llcorner \{(x,S): \psi(x) > 0\}$ *is integral, and*

(c) $\overline{D}\|V_t\|(\psi) \le \delta(V_t,\psi)(\underline{\underline{h}}(V_t,\cdot))$.

Proof: Suppose $0 < r < s$. For any $\tau > 0$, 4.14(b) implies that $\overline{D}\|V_\lambda\|(\psi + \tau\Omega)$ has a finite upper bound for $0 \le \lambda \le s$. Since 4.27 applies for almost all $\lambda > 0$, we may calculate

(1) $$\|V_s\|(\psi) - \|V_r\|(\psi)$$

$$= \lim_{\tau\to 0^+}\{\|V_s\|(\psi + \tau\Omega) - \|V_r\|(\psi + \tau\Omega)\}$$

$$\le \liminf_{\tau\to 0^+} \int_r^s \overline{D}\|V_\lambda\|(\psi + \tau\Omega)\,d\lambda$$

$$\le \liminf_{\tau\to 0^+} \int_r^s \delta(V_\lambda, \psi + \tau\Omega)(\underline{\underline{h}}(V_\lambda,\cdot))\,d\lambda$$

$$\le \liminf_{\tau\to 0^+} \int_r^s \delta(V_\lambda,\psi)(\underline{\underline{h}}(V_\lambda,\cdot)) + \tau\delta(V_\lambda,\Omega)(\underline{\underline{h}}(V_\lambda,\cdot))\,d\lambda$$

$$\le \int_r^s \delta(V_\lambda,\psi)(\underline{\underline{h}}(V_\lambda,\cdot))\,d\lambda + \liminf_{\tau\to 0^+} \tau \int_r^s \|V_\lambda\|(\Omega)\,d\lambda$$

$$\le \int_r^s \delta(V_\lambda,\psi)(\underline{\underline{h}}(V_\lambda,\cdot))\,d\lambda,$$

where we have used 3.4 to estimate

$$\delta(V_\lambda,\Omega)(\underline{\underline{h}}(V_\lambda,\cdot)) \le \|V_\lambda\|(\Omega).$$

For each $0 < B < \infty$ let $E_B \subset \underline{\underline{R}}^+$ consist of those $u \in \underline{\underline{R}}^+$ such that V_u is rectifiable and

(2) $$\delta(V_u,\psi)(\underline{\underline{h}}(V_u,\cdot)) > -B.$$

For $\eta > 0$, let

(3) $$W_\eta = \{x \in \underline{\underline{R}}^n: \psi(x) > \eta\}.$$

Then for $0 < u < t$ by definition

(4) $$\|\delta V_u\| W_\eta = \int_{W_\eta} |\underline{\underline{h}}(V_u,x)|\, d\|V_u\| x,$$

so by Schwarz' inequality,

(5) $$\int |\underline{\underline{h}}(V_u,x)|^2 \psi(x)\, d\|V_u\| x$$

$$\geq [\int_{W_\eta} |\underline{\underline{h}}(V_u,x)|\, d\|V_u\| x]^2 / \int_{W_\eta} \psi(x)^{-1} d\|V_u\| x$$

$$\geq \eta [\|\delta V_u\| W_\eta]^2 / \|V_u\| W_\eta .$$

Since ψ has compact support, it follows by judiciously choosing ϕ in 4.14(b) that $\|V_u\| W_\eta$ has a finite bound for $0 \leq u \leq t$. Referring back to 3.4, we can conclude that for each $0 < B < \infty$ there is a $0 < C(B) < \infty$ such that if $\|\delta V_u\| W_\eta > C(B)$ then

(6) $$\delta(V_u,\psi)(\underline{\underline{h}}(V_u,\cdot)) < -B.$$

Suppose $D^-\|V_t\|(\psi) > -\infty$. By 3.4,

(7) $$\delta(V_u,\psi)(\underline{\underline{h}}(V_u,\cdot)) \leq \|V_u\|(|D\psi|^2/\psi)$$

and since $|D\psi|^2/\psi$ is bounded with compact support, 4.14(b) implies that there is a finite upper bound K for $\|V_u\|(|D\psi|^2/\psi)$ for $0 < u < t$. Hence for any $0 < B < \infty$ we see from (1) for $0 < s < t$

$$\text{(8)} \qquad \|V_t\|(\psi) - \|V_s\|(\psi) \le \int_{[s,t]-E_B} -B \, d\lambda + \int_s^t K d\lambda$$

$$\le -B\mathscr{L}^1([s,t] \sim E_B) + K(t - s).$$

Hence

$$\text{(9)} \qquad \limsup_{s\to t^-} \frac{\mathscr{L}^1([s,t] \sim E_B)}{t - s} \le B^{-1}[D^-\|V_t\|(\psi) - K].$$

Suppose $u_1, u_2, \ldots \in (0,t) \cap E_B$ and $\lim_{i\to\infty} u_i = t$. By 4.14(e), $\lim_{i\to\infty} \|V_{u_i}\| \mathbin{|} \psi = \|V_t\| \mathbin{|} \psi$. Since $\|\delta V_{u_i}\| W_\eta < C(B)$, [AW1 5.6] would imply that $[\lim_{i\to\infty} V_{u_i}] \llcorner W_\eta$ is rectifiable, so

$$\text{(10)} \qquad [\lim_{i\to\infty} V_{u_i}] \llcorner W_\eta$$

and $V_t \llcorner W_\eta$ would be rectifiable. Now (9) implies that there is some $0 < B < \infty$ such that $[s,t] \cap E_B$ is nonempty for all $s < t$, so $V_t \llcorner W_\eta$ is rectifiable.

Similarly, if V_0 is integral, [AW1 6.5] in place of [AW1 5.6] implies $V_t \llcorner W_\eta$ is integral. The same basic arguments hold if $D^+\|V_t\|(\psi) > -\infty$. Since η was arbitrary, conclusions (a) and (b) follow.

In regard to conclusion (c), if $\overline{D}\|V_t\|(\psi) = -\infty$, we are done. Suppose $D^-\|V_t\|(\psi) > -\infty$, and let $\eta > 0$. It follows from the first part of this proof and 4.28 that there is $0 < B < \infty$ and $0 < s < t$ such that whenever $s < u < t$ then

$$\text{(11)} \qquad \mathscr{L}^1([u,t] \sim E_B) < (t-u)\eta/2K,$$

and if $\lambda \in [s,t] \cap E_B$ then

$$\text{(12)} \qquad \delta(V_\lambda,\psi)(\underline{\underline{h}}(V_\lambda,\cdot)) \leq \delta(V_t,\psi)(\underline{\underline{h}}(V_t,\cdot)) + \eta/2.$$

Using (1), (7), (11), (12) and the definition of K,

$$\begin{aligned} &\|V_t\|(\psi) - \|V_u\|(\psi) \\ &\leq \int_{[u,t]\cap E_B} \delta(V_t,\psi)(\underline{\underline{h}}(V_t,\cdot)) + \eta/2 \, d\lambda + \int_{[u,t]-E_B} K \, d\lambda \\ &\leq (t-u)[\delta(V_t,\psi)(\underline{\underline{h}}(V_t,\cdot)) + \eta/2] + (t-u)(\varepsilon/2K)K \\ &\leq (t-u)[\delta(V_t,\psi)(\underline{\underline{h}}(V_t,\cdot)) + \eta]. \end{aligned}$$

Hence

$$D^-\|V_t\|(\psi) \leq \delta(V_t,\psi)(\underline{\underline{h}}(V_t,\cdot)) + \eta.$$

Since η was arbitrary, conclusion (c) follows. The same argument works if $D^+\|V_t\|(\psi) > -\infty$. □

5. Perpendicularity of mean curvature.

We shall show in this chapter that if V is an integral varifold and $\|\delta V\|$ is a Radon measure, then the mean curvature vector $\underline{h}(V,x)$ is perpendicular to the varifold at $\|V\|$ almost all x. This says nothing about singular first variation, but there will be no singular first variation present in our applications in chapter 6.

One may think of the mean curvature vector as pointing in the direction of increasing mass. On a smooth manifold, mass does not increase in any tangential direction because of the local flatness; hence the mean curvature vector is perpendicular to the manifold. We shall show that the varifolds under study have a certain amount of local flatness, and then the integral density hypothesis will imply that there is very little tangential variation in mass.

By definition, integral varifolds are locally flat in the sense that they have approximate tangent planes almost everywhere. But this is not quite flat enough. Therefore we have adapted the method of [AW1 chap. 8]: first we show that a nearly flat piece of varifold can be approximated with a Lipschitz afunction, and then we show that this function is nearly harmonic if the first variation is not too badly behaved. Well-known properties of harmonic functions give the desired additional flatness.

The Lipschitz approximation theorem 5.4 will be used frequently in chapter 6 and promises wide application in future studies. Therefore it is proved in fairly broad generality.

5.1. Definitions.

Let $\underline{\underline{A}}(n,k)$ be the family of affine subsets $S + a$ corresponding to $S \in \underline{\underline{G}}(n,k)$ and $a \in \underline{\underline{R}}^n$.

Let $\chi\colon \underline{\underline{R}}^n \to \underline{\underline{R}}^+$ be an infinitely differentiable function such that $\chi(x)$ is a decreasing function of $|x|$, $\operatorname{spt} \chi \subset \underline{\underline{B}}(0,1)$, and $\chi(x) = 1$ when $|x| \leq 1 - 1/100k$. If $0 < R < \infty$, define $\chi(R,x) = \chi(x/R)$. If $T \in \underline{\underline{G}}(n,k)$, define $\chi_T(R,x) = \chi(T(x)/R)$. Set

$$\rho = \sup_{x \in \underline{\underline{R}}^n} \{|D\chi(x)|,\ \|D^2\chi(x)\|,\ |D\chi(x)|^2/\chi(x)\},$$

$$\underline{\underline{\beta}} = \int_T \chi^2(x)\, d\mathscr{H}^k x.$$

Note that $\underline{\underline{\beta}} > (99/100)\underline{\underline{\alpha}}$. We will often use χ^2 as an approximation to the characteristic function of the unit ball. The square is technically convenient.

In several of the following theorems, there will occur the expression

(1) $$\int |\underline{\underline{h}}(V,x)|^p \phi(x)\, d\|V\|x$$

where $1 \leq p < \infty$ and ϕ is nonnegative. We extend the meaning of (1) when $\|\delta V\|_{\text{sing}}(\phi) > 0$ by

$$\int |\underline{\underline{h}}(V,x)|^p \phi(x)\, d\|V\|x = \begin{cases} \|\delta V\|(\phi) & \text{when } p = 1 \\ \infty & \text{when } p > 1. \end{cases}$$

5.2. Multiple valued Lipschitz functions.

A nearly flat piece of an integral varifold may be essentially multi-layered. To approximate varifolds with this behavior, we shall consider Lipschitz functions $f: \mathbb{R}^k \to \mathbb{M}_\nu$, where $\nu \in \mathbb{N}$ and $\mathbb{M}_\nu$ is the quotient space of $(\mathbb{R}^{n-k})^\nu$ under the equivalence relation $(w_1,\ldots,w_\nu) \equiv (z_1,\ldots,z_\nu)$ if and only if $(w_1,\ldots,w_\nu)$ is a permutation of $(z_1,\ldots,z_\nu)$. For $y \in \mathbb{R}^k$, we let $(f(y)_1,\ldots,f(y)_\nu)$ be any representative of $f(y)$, and if f is differentiable at y, then we must understand $Df(y)_j$ in the sense of $[Df(y)]_j$ rather than $D[f(y)_j]$.

If we define $F: \mathbb{R}^k \to \mathbb{R}^k \times \mathbb{M}_\nu$ by $F(y) = (y,f(y))$, then we also define

$$DF(y)_j = Dy \oplus Df(y)_j \quad \text{for} \quad j = 1,\ldots,\nu.$$

We further define

$$\star\text{-image } F = \{x \in \mathbb{R}^n: x = F(y)_j \text{ for some } y \in \mathbb{R}^k \text{ and some } j = 1,\ldots,\nu\}.$$

The quotient metric on $\mathbb{M}_\nu$ is

$$|w-z| = \inf_{\pi\in\Pi} \Big(\sum_{i=1}^{\nu} |w_i - z_{\pi(i)}|^2\Big)^{1/2}$$

where Π is the set of permutations of ν elements. There is a bi-Lipschitz imbedding of $\mathbb{M}_\nu$ in a higher dimension Euclidean space, and the image of $\underline{\underline{M}}_\nu$ is a Lipschitz retract of the whole space. Therefore Kirszbraun's theorem [FH 2.10.43] on the extension of Lipschitz functions applies to $\underline{\underline{M}}_\nu$, but the Lipschitz constant of the extension may be greater than that of the original map by some factor $c(\nu)$.

5.3. Multilayer monotonicity.

This lemma shows that if a nearly horizontal varifold passes through ν vertically separated points and has small first variation, then the varifold has at least ν layers in a neighborhood of those points.

Lemma: Corresponding to each λ, ξ, and ν such that $0 < \lambda < 1$, $1 \leq \xi < \infty$, and $\nu \in \underline{\underline{N}}$, there is $\gamma > 0$ with the following property:

If $V \in \underline{\underline{IV}}_k(\underline{\underline{R}}^n)$, $Y \subset \underline{\underline{R}}^n$, $\operatorname{card} Y \leq \nu$, $T \in \underline{\underline{G}}(n,k)$, $0 < R < \infty$, $b \in T$, $|b| \leq R$,

(1) $\quad |y - z| \leq \xi |T^{\perp}(y - z)|$ whenever $y,z \in Y$,

(2) $\quad \Theta^k(\|V\|,y) \in \underline{\underline{N}}$ for $y \in Y$,

(3) $\quad \sum\{\Theta^k(\|V\|,y) : y \in Y\} \geq \nu$,

(4) $\quad \int_{\underline{\underline{B}}(y,r)} \|S-T\| dV(x,S) \leq \gamma \|V\| \underline{\underline{B}}(y,r)$

whenever $0 < r \le R + |b|$, $y \in Y$, *and* $b \neq 0$ *or* $\nu \ge 2$, *and*

(5) $$r\|\delta V\|\underline{\underline{B}}(y,r) \le \gamma\|V\|\underline{\underline{B}}(y,r)$$

whenever $0 < r \le R + |b|$ *and* $y \in Y$, *then*

(6) $$\|V\|\{x: \text{dist}(x-b,Y) < R\} \ge \lambda\nu\underline{\underline{\alpha}}R^k.$$

Proof: Because of the behavior of the various quantities under homothety and translation, we may assume $T = \underline{\underline{e}}_1 \wedge \ldots \wedge \underline{\underline{e}}_k$ and $R = 1$.

Suppose the lemma were not true. Then for each $m \in \underline{\underline{N}}$ there would be $V_m \in \underline{\underline{IV}}_k(\underline{\underline{R}}^n)$, $Y_m \in \underline{\underline{R}}^n$, and $k_m \in T$ satisfying $|b_m| \le 1$,

(7) $$|y - z| \le \xi|T^{\perp}(y - z)| \quad \text{whenever} \quad y,z \in Y_m,$$

(8) $$\Theta^k(\|V_m\|,y) \in \underline{\underline{N}} \quad \text{for} \quad y \in Y_m,$$

(9) $$\text{card } Y \le \nu, \quad \sum\{\Theta^k(\|V_m\|,y): y \in Y_m\} \ge \nu,$$

(10) $$\int_{\underline{\underline{B}}(y,r)} \|S-T\| dV_m(x,S) < (1/m)\|V_m\|\underline{\underline{B}}(y,r)$$

whenever $0 < r \le 1 + |b_m|$, $y \in Y_m$, and $b_m \neq 0$ or $\nu \ge 2$,

(11) $$r\|\delta V_m\|\underline{\underline{B}}(y,r) \le (1/m)\|V_m\|\underline{\underline{B}}(y,r)$$

whenever $0 < r \le 1 + |b_m|$ and $y \in Y_m$, and

(12) $$\|V_m\|\{x: \operatorname{dist}(x-b_m, Y_m) < 1\} < \lambda\nu\underline{\underline{\alpha}}R^k.$$

Define R_m to be the supremum of those $r < 1$ for which

$$\|V_m\|\{x: \operatorname{dist}(x-sb_m, Y_m) \leq s\} \geq \lambda\nu\underline{\underline{\alpha}}s^k$$

for $0 < s \leq r$. Condition (9) guarantees $R_m > 0$. Now let

$$W_m = \underline{\underline{\mu}}(1/R_m)_{\#}V_m, \quad Y_m^* = R_m^{-1}Y_m, \quad \text{and}$$

$$A_m = \{x: \operatorname{dist}(x-b_m, Y_m^*) < 1\}.$$

We will be concerned only with $V_m \llcorner A_m$, so by cutting out and moving around chunks of varifold, we may assume that there is some bounded set containing every A_m.

It follows from the definition of R_m that

(13) $$\|W_m\|A_m = \lambda\nu\underline{\underline{\alpha}},$$

and (11) implies that

(14) $$\lim_{m\to\infty}\|\delta W_m\|A_m = 0.$$

To each Y_m^* associate a ν-tuple $Z_m = (y_{m1}, \ldots, y_{m\nu})$ such that $\{y_{m1}, \ldots, y_{m\nu}\} = Y_m^*$ and

$$\operatorname{card}\{j: y_{mi} = y_{mj}\} \leq \Theta^k(\|W_m\|, y_i)$$

for $i = 1,\ldots,\nu$. Then, by the compactness theorem for integral varifolds [AW1 6.4] and the compactness properties of Euclidean spaces, there are convergent subsequences (labelled the same)

$$W_m \llcorner A_m \to W \in \mathbf{IV}_k(\mathbf{R}^n),$$

$$b_m \to b, \quad Z_m \to Z, \quad A_m \to A, \quad Y_m^* \to Y.$$

It follows from the definition of R_m that

$$\|W\|\{x\colon \operatorname{dist}(x-sb,Y) < s\} \geq \lambda\nu\boldsymbol{\alpha}s^k \tag{15}$$

for $0 < s \leq 1$. By (14), W is stationary in A.

In case $\nu = 1$, $b = 0$, and $Y = \{y\}$, we see that $y \in \operatorname{spt}\|W\|$. Hence $\Theta^k(\|W\|,y) \geq 1$ by the upper semicontinuity of density for stationary varifolds [AW1 8.6]. The monotonicity lemma 4.17 yields $\|W\|\mathbf{U}(y,1) \geq \boldsymbol{\alpha}$, which contradicts (13).

Otherwise, it follows from (10) that $S = T$ for W almost all $(x,S) \in \mathbf{G}_k(\mathbf{R}^n)$. Being stationary in A, W must be of the form

$$W = \sum_j [q_j \mathbf{v}(T+d_j) \llcorner A], \tag{16}$$

where $q_j \in \{0\} \cup \mathbf{N}$ and $d_j \in \mathbf{R}^n$. We may suppose $Y = \{d_1,\ldots,d_\nu\}$. By (7), if $i,j \leq \nu$ and $d_i \neq d_j$, then $|T^\perp(d_i-d_j)| > 0$. Hence (15) implies $\sum_{j=1}^{\nu} q_j \geq \lambda\nu$. But since W is integral, it must be true that $\sum_{j=1}^{\nu} q_j \geq \nu$. Then (16) says that $\|W\|A \geq \nu\boldsymbol{\alpha}$, which contradicts (13) again. □

Remark: The essential difference between this lemma and [AW1 6.2] is that hypothesis (5) involves $r\|\delta V\|\underline{\underline{B}}(y,r)$ instead of $R\|\delta V\|\underline{\underline{B}}(y,r)$. This seemingly slight difference is actually the key to chapter 6, for it gives rise to $\alpha^{pk/(k-p)}$ in 5.4(10) instead of α^p, and this in turn permits things to happen in finite time in 6.7. The proof shows why we are assuming a discrete range of values for $\Theta^k(\|V\|,x)$ rather than a continuous range bounded below away from zero.

5.4. Lipschitz approximation.

Theorem: For each p, ν, and ε with $1 \le p < \infty$, $\nu \in \underline{\underline{N}}$, and $0 < \varepsilon < 1$, there exists P with $0 < P < \infty$ such that if

(1) $$V \in \underline{\underline{IV}}_k(\underline{\underline{R}}^n), \quad T = \underline{\underline{e}}_1 \wedge \ldots \wedge \underline{\underline{e}}_k \in \underline{\underline{G}}(n,k),$$

(2) $$(\nu-1+\varepsilon)\underline{\underline{\alpha}} < \|V\|\underline{\underline{B}}(0,1),$$

$$\|V\|\underline{\underline{B}}(0,3) \le (\nu+1-\varepsilon)\underline{\underline{\alpha}}3^k,$$

(3) $$\alpha^p = \int_{\underline{\underline{B}}(0,7)} |\underline{\underline{h}}(V,x)|^p d\|V\|x,$$

(4) $$\beta^2 = \int_{\underline{\underline{B}}(0,7)} \|S-T\|^2 dV(x,S),$$

(5) $$1 \le q < \infty \text{ and } \mu^q = \int_{\underline{\underline{B}}(0,7)} |\underset{\sim}{T}^{\perp}(x)|^q d\|V\|x,$$

then there are Lipschitz maps $f: T \to \underline{\underline{M}}_\nu$ and $F: T \to T \times \underline{\underline{M}}_\nu$ such that

(6) $$F(z) = (z, f(z)) \quad \underline{\text{for}} \quad z \in T,$$

(7) $$|f(z_1) - f(z_2)| \leq c(\nu)|z_1 - z_2| \quad \underline{\text{for}} \quad z_1, z_2 \in T,$$

(8) $$\sup\{|f(z)_i| : z \in T, \quad i = 1, \ldots, \nu\} \leq 4(\mu^q/\underline{\alpha})^{1/(k+q)},$$

and if

(9) $$Y = \{z \in \mathbf{B}^k(0,1): F(z_i) \in \mathbf{B}(0,1) \quad \text{and}$$

$$\theta^k(\|V\|, F(z)_i) = \text{card}\{j: F(z)_i = F(z)_j\} \quad \text{for} \quad i = 1, \ldots, \nu\},$$

$$X = \mathbf{B}(0,1) \cap \text{*-image } F \cap T^{-1}(Y)$$

then

(10) $$\|V\|(\mathbf{B}(0,1) \sim X) + \mathscr{H}^k(\mathbf{B}^k(0,1) \sim Y)$$

$$\leq \begin{cases} P[\alpha^{pk/(k-p)} + \beta^2 + \mu^q] & \underline{\text{if}} \quad p < k \\ P[\beta^2 + \mu^q] & \underline{\text{if}} \quad p \geq k. \end{cases}$$

Proof: The basic idea of the proof is that the points to which we can apply the multilayer monotonicity lemma 5.3 cannot stack up more than ν deep and are related in a Lipschitzian manner in horizontal directions.

Let $2/3 < \lambda < 1$ and $0 < \gamma < 1$ be such that

(11) $$\lambda(\nu+1)(3-\gamma)^k > (\nu+1-\varepsilon)3^k,$$

$$\lambda\nu(2-\gamma)^k \geq \nu 2^k - 4^{-k-1},$$

and γ works in 5.3 for λ and $\nu + 1$ with $\xi = 2^{1/2}$. Let A be the set of those $y \in \mathbf{B}(0,2)$ such that $\Theta^k(\|V\|,y) \in \mathbf{N}$, $|T^{\perp}(y)| < \gamma/4$, and

(12) $$r\|\delta V\|\mathbf{B}(y,r) < \gamma\|V\|\mathbf{B}(y,r) \quad \text{and}$$

(13) $$\int_{\mathbf{B}(y,r)} \|S-T\|^2 dV(x,S) < \gamma^2\|V\|\mathbf{B}(y,r)$$

whenever $0 < r \leq 5$. Define

(14) $$B = \{x \in \mathbf{B}(0,2) \sim A\colon \Theta^k(\|V\|,x) \in \mathbf{N}\},$$

$$C = T(B).$$

Suppose $z \in \mathbf{B}^k(0,2) \sim C$. Then for $y \in A$, $T(y) = z$, and $0 < r \leq 5$ we have from (13) and Schwarz' inequality,

(14) $$\int_{\mathbf{B}(y,r)} \|S-T\| dV(x,S) < \gamma\|V\|\mathbf{B}(y,r).$$

Applying the multilayer monotonicity lemma 5.3 with $b = z$ and $R = 3 - \gamma$ shows that if

(15) $$\sum\{\Theta^k(\|V\|,y)\colon y \in A,\ T(y) = z\} \leq \nu$$

does not hold, then

(16) $$\|V\|\underline{B}(0,3) > \lambda(\nu+1)\underline{\alpha}(3-\gamma)^k.$$

By the choice of λ and γ in (11), (16) would contradict hypothesis (2). Therefore (15) does hold.

Next, we put a bound on $|T^{\perp}(y)|$ for $y \in A$. Let $\sigma = 4[\mu^q/\underline{\alpha}]^{1/(k+q)}$ and suppose that $|T^{\perp}(y)| > \sigma$ for some $y \in A$. Then by the monotonicity lemma 5.3 we get

$$\|V\|\underline{B}(y,\sigma/2) \geq (\underline{\alpha}/2)(\sigma/2)^k,$$

which certainly means that

$$\int_{\underline{B}(0,6)} |T^{\perp}(x)|^q d\|V\|x \geq (\underline{\alpha}/2)(\sigma/2)^{k+q}$$
$$\geq 2^{k+q-1}\mu^q,$$

which contradicts the definition of μ. Hence

(17) $$\sup\{|T^{\perp}(y)| : y \in A\} \leq 4[\mu^q/\underline{\alpha}]^{1/(k+q)}.$$

Now define the set E to consist of those $z \in \underline{B}^k(0,2) \sim C$ such that

$$\sum\{\Theta^k(\|V\|,y) : y \in A,\ T(y) = z\} = \nu$$

and define $f: E \to \underline{M}_\nu$ and $F: E \to T \times \underline{M}_\nu$ so that $F(z)_i = (z, f(z)_i) \in A$ and

$$\operatorname{card}\{j: f(z)_j = f(z)_i\} = \Theta^k(\|V\|, f(z)_i)$$

for $i = 1,\ldots,\nu$ and $z \in E$. To see that F is Lipschitz, suppose $z_1, z_2 \in E$. If $|z_1 - z_2| < \gamma/2$, then the components of $F(z_1)$ and $F(z_2)$ can be paired off so that $|F(z_1)_i - F(z_2)_i| < 2^{1/2}|z_1 - z_2|$, or else we could pick $\nu + 1$ points from $F(z_1) \cup F(z_2)$ and apply 5.3 to get a contradiction to (3). Thus

$$\text{(18)} \qquad |f(z_1) - f(z_2)| \leq [\sum_i |f(z_1)_i - f(z_2)_i|^2]^{1/2}$$

$$\leq 2\nu^{1/2}|z_1 - z_2|.$$

Since $|f(z)_i| < \gamma/4$ holds by the definition of A, we see that (18) also holds if $|z_1 - z_2| > \gamma/2$, and so f has Lipschitz constant $2\nu^{1/2}$ on E. We then use Kirszbraun's Theorem as noted in 5.2 to obtain Lipschitz extensions $f: T \to \underline{\underline{M}}_\nu$ and $F: T \to T \times \underline{\underline{M}}_\nu$ satisfying (6), (7), and (8).

The rest of the proof verifies (10). First, we estimate $\|V\|B$. Suppose $b \in B$. If it is (12) that fails for b, then we can choose $r(b)$ such that $0 < r(b) < 5$,

$$\text{(19)} \qquad \|V\|\underline{\underline{B}}(b, r(b)) \geq (1/2)\underline{\underline{\alpha}}r(b)^k \quad \text{and}$$

$$\text{(20)} \qquad r(b)\|\delta V\|\underline{\underline{B}}(b, r(b)) \geq \gamma\|V\|\underline{\underline{B}}(b, r(b)),$$

either by choosing the smallest $r(b)$ for which (20) holds and using monotonicity lemma 5.5 to get (19), or otherwise using

$\Theta^k(\|V\|,b) \geq 1$ to get (19) and (20) to hold for the same r. Hölder's inequality applied to (20) yields

$$r(b)^p \int_{\underline{\underline{B}}(b,r(b))} |\underline{\underline{h}}(V,x)|^p d\|V\|x \geq \gamma^p \|V\| \underline{\underline{B}}(b,r(b)),$$

so, using (19),

$$\int_{\underline{\underline{B}}(b,r(b))} |\underline{\underline{h}}(V,x)|^p d\|V\|x \geq (\gamma^p/2)\underline{\underline{\alpha}} r(b)^{k-p}. \tag{21}$$

If $p \geq k$, then P can be chosen large enough so that either (10) holds trivially or else α must be so small that (21) cannot hold and B is empty. Otherwise, if $p < k$,

$$r(b) \leq [2\alpha^p/\gamma^p \underline{\underline{\alpha}}]^{1/(k-p)}$$

and (20) may be replaced by

$$\int_{\underline{\underline{B}}(b,r(b))} |\underline{\underline{h}}(V,x)|^p d\|V\|x \geq \gamma^p [\gamma^p \underline{\underline{\alpha}}/2\alpha^p]^{p/(k-p)} \|V\| \underline{\underline{B}}(b,r(b)). \tag{22}$$

If (12) does hold for b, then either $|T^{\perp}(b)| > \gamma/4$ or there is some $r(b)$ with $0 < r(b) \leq 5$ and

$$\int_{\underline{\underline{B}}(b,r(b))} \|S-T\|^2 dV(x,S) \geq \gamma^2 \|V\| \underline{\underline{B}}(b,r(b)). \tag{23}$$

Hence the Besicovitch covering theorem 2.2 implies

$$(24) \qquad \|V\|B \le (4/\gamma)^q \int_{\underline{\underline{B}}(0,2)} |T^{\perp}(x)|^q d\|V\|x$$

$$+ \underline{\underline{B}}(n)\gamma^{-p}[2\alpha^p/\gamma^p\alpha]^{p/(k-p)} \int_{\underline{\underline{B}}(0,7)} |\underline{\underline{h}}(V,x)|^p d\|V\|x$$

$$+ \underline{\underline{B}}(n)\gamma^{-2} \int_{\underline{\underline{B}}(0,7)} \|S-T\|^2 dV(x,S).$$

$$\le (4/\gamma)^q \mu^q + \underline{\underline{B}}(n)\gamma^{-kp/(k-p)}(2/\underline{\underline{\alpha}})^{p/(k-p)}\alpha^{kp/(k-p)}$$

$$+ \underline{\underline{B}}(n)\gamma^{-2}\beta^2,$$

but recall that the α term is absent if $p \ge k$.

Our next aim is to find out how much of $\underline{\underline{B}}^k(0,2)$ is covered by less than ν layers of A. Recall that E is where A has ν layers, and let $Q = \underline{\underline{B}}^k(0,2) \sim E$. When G is any $\mathscr{H}^k$ measurable subset of $\underline{\underline{B}}^k(0,2)$,

$$\|V\|[\underline{\underline{B}}(0,2) \cap T^{-1}(G)] \le \|V\|[A \cap T^{-1}(E \cap G)]$$

$$+ \|V\|[A \cap T^{-1}(Q \cap G)] + \|V\|B.$$

Since A has no more than ν layers,

$$\nu\mathscr{H}^k(E \cap G) \ge \int_{A\cap T^{-1}(E\cap G)} |\Lambda_k T\circ S|\, dV(x,S)$$

$$\ge \int_{A\cap T^{-1}(E\cap G)} 1 - k\|S-T\|^2 dV(x,S).$$

Likewise, since A has no more than $\nu - 1$ layers over Q,

$$(25)\qquad (\nu-1)\mathcal{H}^k(Q \cap G) \geq \int_{A\cap T^{-1}(Q\cap G)} 1 - k\|S-T\|^2 dV(x,S).$$

Hence

$$(26)\qquad \|V\|[\underline{\underline{B}}(0,2) \cap T^{-1}(G)] \leq \nu\mathcal{H}^k(E \cap G) + (\nu - 1)\mathcal{H}^k(Q \cap G)$$

$$+ k\int_{\underline{\underline{B}}(0,2)} \|S-T\|^2 dV(x,S) + \|V\|B.$$

Now consider $G = \underline{\underline{B}}^k(0,1)$. We see from (24) that $\|V\|B$ is small if α, β, and μ are small. Thus we can pick P large enough so that either (10) holds trivially, or else α, β, and μ are small enough so that (2) and (26) imply that $E \cap \underline{\underline{B}}(0,1)$ is nonempty. Then by the multilayer monotonicity theorem and (11)

$$\|V\|\underline{\underline{B}}(0,2) \geq \lambda\nu\underline{\underline{\alpha}}(2-\gamma)^k$$

$$\geq \nu\underline{\underline{\alpha}}2^k - 4^{-k-1}\underline{\underline{\alpha}}.$$

Hence, using (26) with $G = \underline{\underline{B}}^k(0,2)$,

$$\nu\underline{\underline{\alpha}}2^k - 4^{-k-1}\underline{\underline{\alpha}} < \nu\mathcal{H}^k(G) - \mathcal{H}^k(Q) + k\beta^2 + \|V\|B.$$

Thus, for large enough P, we may assume

$$\mathcal{H}^k(Q) < 4^{-k}\underline{\underline{\alpha}}.$$

Next, let $Q^* = Q \cap \underline{\underline{B}}^k(0,1)$. Since $\mathcal{H}^k$ almost every point of Q^* is a Lebesgue point of Q, and since $\mathcal{H}^k(Q) < 4^{-k}\underline{\underline{\alpha}}$, we can for $\mathscr{H}^k$ almost every $w \in Q^*$ choose $r(w)$ with $0 < r(w) < 1$ and

(27) $$\mathscr{H}^k[Q \cap \underline{\underline{B}}^k(w,r(w))] = 4^{-k}\underline{\underline{\alpha}}r(w)^k.$$

By the Besicovitch covering theorem 2.2, there is a collection $\mathscr{B}$ of disjoint balls $\underline{\underline{B}}(w,r(w))$ with $w \in Q^*$ and

(28) $$\mathscr{H}^k(Q \cap \cup \mathscr{B}) \geq \mathscr{H}^k(Q^*)/\underline{\underline{B}}(k).$$

Using $G = \cup\mathscr{B}$ in (26) produces

$$\|V\|(\underline{\underline{B}}(0,2) \cap T^{-1}(\cup\mathscr{B})) \leq \nu\mathscr{H}^k(E \cap \cup \mathscr{B})$$

$$+ (\nu-1)\mathscr{H}^k(Q \cap \cup \mathscr{B}) + k\beta^2 + \|V\|B.$$

But the condition (27) guarantees that in each $\underline{\underline{B}}(w,r(w))$ there is $x(w) \in E$ with $|x(w) - w| < r(w)/2$. Hence the multilayer monotonicity theorem 5.3 implies that

$$\|V\|\{y \in \underline{\underline{B}}(0,2): |T(y) - w| \leq r(w)\} \geq \lambda\nu\underline{\underline{\alpha}}r(w)^k$$

and so

$$\|V\|(\underline{\underline{B}}(0,2) \cap T^{-1}(\cup\mathscr{B})) > \lambda\nu\mathscr{H}^k(\cup\mathscr{B}).$$

Thus, using (26) with $G = \cup\mathscr{B}$,

(29) $$\lambda\nu\mathscr{H}^k(\cup\mathscr{B}) \leq \nu\mathscr{H}^k(\cup\mathscr{B}) - \mathscr{H}^k(Q \cap \cup\mathscr{B})$$

$$+ k\beta^2 + \|V\|B.$$

It follows from (26) that

$$\mathscr{H}^k(\cup\mathscr{B}) = 4^k\mathscr{H}^k(Q \cap \cup\mathscr{B}),$$

and so (28) becomes

$$\mathscr{H}^k(Q \cap \cup\mathscr{B}) \leq (1-\lambda)\nu 4^k\mathscr{H}^k(Q \cap \cup\mathscr{B}) + k\beta^2 + \|V\|B.$$

Using the properties of λ,

$$\mathscr{H}^k(Q \cap \cup\mathscr{B}) \leq 2(k\beta^2 + \|V\|B).$$

Thence, by (28),

(30) $$\mathscr{H}^k(Q^*) \leq 2\underline{\underline{B}}(k)(k\beta^2 + \|V\|B).$$

Now to verify (10). We have

$$\underline{\underline{B}}(0,1) \sim X \subset B \cup [A \cap T^{-1}(Q^*)],$$

so, using (25) with $G = Q^*$,

(31) $$\|V\|[\underline{\underline{B}}(0,1) \sim X] < \|V\|B + (\nu-1)\mathscr{H}^k(Q^*) + k\beta^2.$$

Since $\underline{\underline{B}}^k(0,1) \sim Y = Q^*$, we see from (24), (30), (31) and the earlier constraints placed on P that there is $P < \infty$ such that

$$\|V\|[\mathbf{B}(0,1) \sim X] + \mathcal{H}^k(Q^*) \le \begin{cases} P[\alpha^{pk/(k-p)} + \beta^2 + \mu^q] & \text{if } p < k \\ P[\beta^2 + \mu^q] & \text{if } p \ge k, \end{cases}$$

which verifies conclusion (10). □

5.5. Tilt of tangent planes.

Here we estimate the total tilt of the tangent planes of a varifold near a k-plane in terms of more convenient quantities.

Lemma: If $V \in \mathbf{IV}_k(\mathbf{R}^n)$, $T \in \mathbf{G}(n,k)$, $\phi \in \mathbf{C}_0^1(\mathbf{R}^n,\mathbf{R}^+)$, $p = 1$ or $p = 2$,

(1) $$\alpha^p = \int |\mathbf{h}(V,x)|^p \phi^2(x)\, d\|V\|x,$$

(2) $$\mu^2 = \int |T^\perp(x)|^2 \phi^2(x)\, d\|V\|x,$$

(3) $$\xi^2 = \int |T^\perp(x)|^2 |D\phi(x)|^2 d\|V\|x, \quad \text{and}$$

(4) $$\beta^2 = \int \|S-T\|^2 \phi^2(x)\, dV(x,S),$$

then

(5) $$\beta^2 \le 4k\alpha^{2/3}\mu^{2/3} + 16\xi^2 \quad \text{if } p = 1,$$

(6) $$\beta^2 \le 2\alpha\mu + 16\xi^2 \quad \text{if } p = 2.$$

Proof: Let $g(x) = \phi(x)^2 T^{\perp}(x)$ for $x \in \underline{\underline{R}}^n$. Then for $S \in \underline{\underline{G}}(n,k)$ we have

$$Dg(x)\cdot S = 2\phi(x)S(T^{\perp}(x))\cdot D\phi(x) + \phi(x)^2 T^{\perp}\cdot S$$

and hence

$$(\phi(x)\|S-T\|)^2 \leq \phi(x)^2 T^{\perp}\cdot S$$

$$\leq Dg(x)\cdot S + 2\phi(x)|S(T^{\perp}(x))\cdot D\phi(x)|$$

$$\leq Dg(x)\cdot S + 2\phi(x)\|S-T\||T^{\perp}(x)||D\phi(x)|.$$

Therefore

$$\int \phi(x)^2\|S-T\|^2 dV(x,S)$$

$$\leq |\delta V(g)| + 2\int \phi(x)\|S-T\||T^{\perp}(x)| \; |D\phi(x)|dV(x,S)$$

$$\leq |\delta V(g)| + 2[\int \|S-T\|^2\phi^2(x)dV(x,S)\int |T^{\perp}(x)|^2|D\phi(x)|^2 d\|V\|x]^{1/2}.$$

If $\beta^2 \leq 4\beta\xi$, then

(7) $$\beta^2 \leq 16\xi^2.$$

Otherwise, we must have $\beta^2 \leq 2|\delta V(g)|$. If $p = 2$, then we use Schwarz' inequality:

(8) $$2|\delta V(g)| \leq 2\int |\underline{\underline{h}}(V,x)|\phi^2(x)|T^{\perp}(x)|d\|V\|x$$

$$\leq 2[\int |\underline{\underline{h}}(V,x)|^2\phi^2(x)d\|V\|x\int |T^{\perp}(x)|^2\phi^2(x)d\|V\|x]^{1/2},$$

nd we get conclusion (6) by adding (7) and (8). If $p = 1$, hen we must be more devious. For a temporarily unfixed constant > 0, decompose g into $g_1 + g_2$, where

$$g_1(x) = \begin{cases} \phi^2(x)T^{\perp}(x) & \text{if } |T^{\perp}(x)| \leq 1/M, \\ \phi^2(x)T^{\perp}(x)/M|T^{\perp}(x)| & \text{if } |T^{\perp}(x)| \geq 1/M, \end{cases}$$

$$g_2(x) = \begin{cases} 0 & \text{if } |T^{\perp}(x)| \leq 1/M, \\ \phi^2(x)T^{\perp}(x)(1-1/M|T^{\perp}(x)|) & \text{if } |T^{\perp}(x)| \geq 1/M. \end{cases}$$

ne may calculate that for $S \in \underline{\underline{G}}(n,k)$,

$$\begin{aligned} |Dg_2(x)\cdot S| &\leq 2\phi(x)|D\phi(x)||T^{\perp}(x)| + k\phi^2(x) \\ &\leq |D\phi(x)|^2|T^{\perp}(x)|^2 + (k+1)\phi(x)^2 \\ &\leq |D\phi(x)|^2|T^{\perp}(x)|^2 + (k+1)M^2|T^{\perp}(x)|^2\phi^2(x) \end{aligned}$$

nen $|T^{\perp}(x)| \geq 1/M$. Then

$$\begin{aligned} |\delta V(g)| &\leq |\delta V(g_1)| + |\delta V(g_2)| \\ &\leq (1/M)\|\delta V\|(\phi^2) + \int |Dg_2(x)\cdot S|\, dV(x,S) \\ &\leq \alpha/M + \xi^2 + (k+1)M^2\mu^2. \end{aligned}$$

e value of M that minimizes this expression is $= (\alpha/2(k+1)\mu^2)^{1/3}$. Hence

$$|\delta V(g)| \le (2k+2)^{1/3}\alpha^{2/3}\mu^{2/3} + \xi^2 + 2^{-2/3}(k+1)^{1/3}\alpha^{2/3}\mu^{2/3},$$

which, together with (7), implies conclusion (5). □

5.6. Blowing up and shrinking down.

This is the basic theorem for getting improved flatness. It shows that if there is little mean curvature in a region compared to the bumpiness of the varifold, then a smaller region must be flatter. The basic idea is to blow up the varifold, more vertically than horizontally, to get a harmonic function, rather than just a tangent plane.

Theorem: If $\nu \in \mathbf{N}$ then there exists a constant $c_4 < \infty$ such that:

If $0 < \theta < 1/18$ and $M < \infty$ then there exists $0 < \eta < 1$ with the following property:

If

(1) $V \in \mathbf{IV}_k(\mathbf{R}^n)$, $a \in \mathbf{R}^n$, $0 < R < \infty$, $A \in \mathbf{A}(n,k)$,

(2) $(\nu-1/2)\boldsymbol{\alpha}(R/9)^k \le \|V\|\mathbf{B}(a,R/a)$,

$\|V\|\mathbf{B}(a,R/3) \le (\nu+1/2)\boldsymbol{\alpha}(R/3)^k$,

(3) $\|V\|\{x \in \mathbf{B}(a,R): \Theta^k(\|V\|,x) \ne \nu\} < \eta R^k$,

(4) $\mathrm{dist}(a,A) \le \eta R$,

(5) $$R^{-k-2}\int_{\mathbf{B}(a,R)} \operatorname{dist}(x,A)^2 d\|V\|x = \mu^2 < \eta, \quad \underline{\text{and}}$$

(6) $$R^{-k+1}\|\delta V\|\mathbf{B}(a,R) = \alpha < M\mu^2,$$

then there is $A^* \in \mathbf{A}(n,k)$ such that

(7) $$(\theta R)^{-k-2}\int_{\mathbf{B}(a,\theta R)} \operatorname{dist}(x,A^*)^2 d\|V\|x \le c_4^2\theta^2\mu^2.$$

Proof: Fix ν, Θ, and M and define

(8) $$c_4^2 = 81[2 + 2(c(\nu)+1)^k \nu c_3^2 9^{k+2} k\underline{\alpha}/(k+4)],$$

where c_3 is the constant appearing in (48) and (49) below.

Owing to the behavior of the various quantities appearing in (1)-(7) with respect to transformation by homothetities and Euclidean motions, we see that were the theorem false there would exist $T \in \mathbf{G}(n,k)$ and to each $i \in \mathbf{N}$ there would correspond η_i, V_i, and a_i such that

(9) $$\lim_{i\to\infty} \eta_i = 0,$$

(10) $$V_i \in \mathbf{IV}_k(\mathbf{R}^n), \quad a_i \in T^{-1}(0),$$

(11) $$(\nu-1/2)\underline{\alpha} \le \|V_i\|\mathbf{B}(a_i,1),$$

$$\|V_i\|\mathbf{B}(a_i,3) \le (\nu+1/2)\underline{\alpha}3^k,$$

(12) $$\|V_i\|\{x \in \mathbf{B}(a_i,9) : \Theta^k(\|V_i\|,x) \ne \nu\} < 9^k\eta_i$$

(13) $$|a_i| < 9\eta_i,$$

(14) $$9^{-k-2} \int_{\underline{\underline{B}}(a_i,9)} \operatorname{dist}(x,T)^2 d\|V_i\|x = \mu_i^2 < \eta_i,$$

(15) $$9^{-k+1}\|\delta V\|\underline{\underline{B}}(a_i,9) = \alpha_i < M\mu_i^2,$$

and for every $A^* \in \underline{\underline{G}}(n,k)$

(16) $$(9\theta)^{-k-2} \int_{\underline{\underline{B}}(a_i,9\theta)} \operatorname{dist}(x,A^*)^2 d\|V_i\|x > c_4^2\theta^2\mu_i^2.$$

It is not too hard to see that these conditions imply that

(17) $$\lim_{i\to\infty} a_i = 0 \quad \text{and}$$

(18) $$\lim_{i\to\infty} V_i \llcorner \underline{\underline{U}}(a_i,9) = \nu\underline{\underline{v}}(T \cap \underline{\underline{U}}(0,9)).$$

For each $i \in \underline{\underline{N}}$ we let

(19) $$\beta_i^2 = \int_{\underline{\underline{B}}(0,7)} \|S-T\|^2 dV_i(x,S),$$

and we note that (15) implies $\mu_i > 0$.

We apply the Lipschitz approximation theorem 5.3 with $\varepsilon = 1/2$ to obtain a real number P and mappings $f_i\colon T \to \underline{\underline{M}}_\nu$ such that

(20) $$F_i(y) = (y, f_i(y)) \quad \text{for} \quad y \in T,$$

(21) $$|f_i(y) - f_i(z)| \le c(\nu)|y-z| \quad \text{for} \quad y,z \in T,$$

if $C = \underline{\underline{B}}(0,1)$, $D = \underline{\underline{U}}^k(0,1)$, and

(22) $$X_i = C \cap *\text{-image } F_i \cap T^{-1}(Y_i),$$

$$Y_i = \{y \in D\colon \Theta^k(\|V_i\|, f_i(y)_m) = \text{card}\{j\colon f_i(y)_j = f_i(y)_m$$

$$\text{and } f_i(y)_m \in \underline{\underline{B}}(0,1) \text{ for } m = 1,\ldots,\nu\}$$

then

(23) $$\|V_i\|(C \sim X_i) + \mathscr{H}^k(D \sim Y_i)$$

$$\le \begin{cases} P[(9^k M\mu_i^2)^{k/(k-1)} + \beta_i^2 + \mu_i^2] & \text{if } k > 1, \\ P[\beta_i^2 + \mu_i^2] & \text{if } k = 1, \text{ and} \end{cases}$$

(24) $$\sup\{|f_i(y)_m|\colon y \in T,\ m = 1,\ldots,\nu\}$$

$$< 4(9^{k+2}\mu_i^2/\underline{\underline{\alpha}})^{1/(k+2)}.$$

From (14), (15), and the tilt lemma 5.5 with $\phi = \chi(8,\cdot)$ we calculate

(25) $$\beta_i^2 \le (9^{k-1}M\mu_i^2)^{2/3}(9^{k+2}\mu^2)^{1/3} + 16(\rho/8)^2 9^{k+2}\mu_i^2$$

$$\le [9^k M^{2/3} + 9^{k+2}\rho^2/2]\mu^2.$$

Therefore, there is a positive real number N so that for all $i \in \underline{\underline{N}}$ we have

(26) $\quad \sup\{\beta_i^2, \|V_i\|(C \sim X_i) + \mathscr{H}^k(D \sim Y_i)\} \le N\mu_i^2.$

Since F_i and f_i are Lipschitz, we use Rademacher's theorem [FH 3.1.6] to see that F_i and f_i are differentiable at $\mathscr{H}^k$ alsmot all points $y \in T$, and by (21)

(27) $\quad |Df_i(y)_m| \le c(\nu), |DF_i(y)_m| \le c(\nu) + 1$

for $m = 1,\ldots,\nu$. Moreover, we see from (22) and the integrality of V_i that

(28)
$$\int_{X_i} \zeta(x,S)\,dV_i(x,S)$$
$$= \int_{Y_i} \sum_{m=1}^{\nu} \zeta(F_i(y)_m, \text{ image } DF_i(y)_m)\,|\Lambda_k DF_i(y)_m|\,d\mathscr{H}^k_y$$

whenever ζ is a bounded Baire function on $\underline{\underline{G}}_k(\underline{\underline{R}}^n)$.

We make the following estimates for sufficiently large i: By (28) and the definition of μ_i,

(29)
$$\int_{Y_i} \sum_m |f_i(y)_m|^2 d\mathscr{H}^k y$$
$$\le \int_{Y_i} \sum_m |f_i(y)_m|^2 |\Lambda_k DF_i(y)_m|\,d\mathscr{H}^k y$$
$$\le \int_{X_i} \operatorname{dist}(x,T)^2 d\|V_i\|x$$
$$\le 9^{k+2}\mu_i^2.$$

By (20), (27), (25), and (26),

(30)
$$\int_{Y_i} \sum_m |Df_i(y)_m|^2 d\mathscr{H}^k y$$

$$\le \int_{Y_i} \sum_m |DF_i(y)_m|^2 \|\text{image } DF_i(y)_m - T\|^2 d\mathscr{H}^k y$$

$$\le (c(\nu)+1)^2 \int_{Y_i} \sum_m \|\text{image } DF_i(y)_m - T\|^2 |\Lambda_k DF_i(y)_m| d\mathscr{H}^k y$$

$$\le (c(\nu)+1)^2 \int_{X_i} \|S-T\|^2 dV_i(x,S)$$

$$\le (c(\nu)+1)^2 N\mu_i^2.$$

By (24) and (26),

(31)
$$\int_{D \sim Y_i} \sum_m |f_i(y)_m|^2 d\mathscr{H}^k y$$

$$\le 16\nu [9^{k+2}\mu_i^2/\underline{\alpha}]^{2/(k+2)} N\mu_i^2.$$

By (27) and (26)

(32)
$$\int_{D \sim Y_i} \sum_m |Df_i(y)_m|^2 d\mathscr{H}^k y \le c(\nu)^2 N\mu_i^2.$$

We see from these estimates, (9), and (14) that

(33)
$$\limsup_{i\to\infty} \mu_i^{-2} \int_D (|f_i|^2 + |Df_i|^2) d\mathscr{H}^k < \infty, \quad \text{and}$$

(34)
$$\limsup_{i\to\infty} \mu_i^{-2} \int_D |f_i|^2 d\mathscr{H}^k \le 9^{k+2}.$$

Using the same reasoning that is well known in the case of single valued functions, passing to a subsequence if necessary, we may find an $\underline{\underline{M}}_\nu$ valued $\mathscr{H}^k \llcorner D$ summable function h^* such that

(35) $$\lim_{i\to\infty} \int_D |h^* - \mu_i^{-1} f_i|^2 d\mathscr{H}^k = 0.$$

It follows from (12) that h^* is single-valued, i.e. there is a $T^\perp$ valued $\mathscr{H}^k \llcorner D$ summable function h such that $h^*(y) = (h(y),\ldots,h(y))$. Clearly, by (34),

(36) $$\int_D |h|^2 d\mathscr{H}^k \leq 9^{k+2}.$$

We will now show that h is $\mathscr{H}^k \llcorner D$ almost equal a harmonic function on D. In order to do this, it will suffice to show in view of (35) that for each smooth function $\phi\colon D \to T^\perp$

(37) $$\lim_{i\to\infty} \mu_i^{-1} \int_D \sum_m Df_i(y)_m \cdot D\phi(y)\, d\mathscr{H}^k y = 0.$$

Fixing ϕ, let

(38) $$B = \sup\{|\phi(y)| + |D\phi(y)|\colon y \in D\},$$

(39) $$a_{1,i} = \int_{D\sim Y_i} \sum_m Df_i(y)_m \cdot D\phi(y)\, d\mathscr{H}^k y,$$

(40) $$a_{2,i} = \int_{Y_i} \sum_m Df_i(y)_m \cdot D\phi(y) - [\text{image } DF_i(y)_m \cdot (D\phi(y)\circ T)]\, |\Lambda_k DF_i(y)_m|\, d\mathscr{H}^k y,$$

(41) $$a_{3,i} = \int_{Y_i} \sum_m \text{image } DF_i(y)_m \cdot (D\phi(y)\circ T)\,|\Lambda_k DF_i(y)_m|\,d\mathscr{H}^k y$$

$$- \delta V_i(\phi\circ T), \quad \text{and}$$

(42) $$a_{4,i} = \delta V_i(\phi\circ T).$$

Note that

(43) $$\mu_i^{-1}\int_D \sum_m Df_i(y)_m \cdot D\phi(y)\,d\mathscr{H}^k y = \mu_i^{-1} \sum_{u=1}^{4} a_{j,i}.$$

We now estimate the four quantities $a_{j,i}$ for large i: Using (27) and (26) and (38)

(44) $$|a_{1,i}| \le c(\nu)BN\mu_i^2.$$

Using $\underline{\underline{c}}_2$ as in [AW1 8.14] and (30),

(45) $$|a_{2,i}| \le \underline{\underline{c}}_2 B\int_{Y_i} \sum_m \|Df_i(y)_m\|^2 d\mathscr{H}^k y$$

$$\le \underline{\underline{c}}_2 B(c(\nu)+1)^2 N\mu_i^2.$$

Using (28) and (26),

(46) $$|a_{3,i}| \le \left|\int_{C\sim X_i} D(\phi\circ T)(x)\cdot S\; dV_i(x,S)\right|$$

$$\le B\|V_i\|(C\sim X_i)$$

$$\le BN\mu_i^2.$$

Using (15) and (17),

(47) $$|a_{4,i}| \leq 9^{k+2} MB\mu_i^2.$$

Because $\eta_i \to 0$ and $\mu_i \to 0$, (37) holds and h is harmonic.

As is well known [FH 5.2.5], there is a positive real number c_3, independent of h, such that when $|y| < 1/2$

(48) $$\sup\{|h(0)|, \|Dh(0)\|\} \leq c_3\left(\int_D |h|^2 d\mathscr{H}^k\right)^{1/2} \quad \text{and}$$

(49) $$|h(y)-h(0)-y\cdot Dh(0)| \leq c_3\left(\int_D |h|^2 d\mathscr{H}^k\right)^{1/2} |y|^2.$$

Whenever i is sufficiently large, we let

(50) $$L_i(y) = y + \mu_i y\cdot Dh(0) \qquad \text{for} \quad y \in T,$$

(51) $$K_i(x) = L_i[T(x)] + \mu_i h(0) \qquad \text{for} \quad x \in \underline{\underline{R}}^n,$$

(52) $$A_i^* = \text{image } K_i \in \underline{\underline{A}}(n,k).$$

If $x \in C$ then

(53) $$x-K_i(x) = T^{\perp}(x)-\mu_i h(0)-\mu_i T(x)\cdot Dh(0),$$

so that, using (48) and (36),

(54) $$\begin{aligned} \operatorname{dist}(x,A_i^*) &\leq |x-K_i(x)| \\ &\leq \operatorname{dist}(x,T) + 2c_3 9^{(k+2)/2}\mu_i. \end{aligned}$$

If $y \in Y_i$ and $m = 1,\dots,\nu$ then

(55) $$F_i(y)_m - K_i(F_i(y)_m) = f_i(y)_m - \mu_i h(y) + \mu_i [h(y) - h(0) - y \cdot Dh(0)]$$

so that, using (49) and (36), for $|y| < 1/2$

(56) $$\sum_m \operatorname{dist}(F_i(y)_m, \tilde{A}_i)^2$$
$$\leq 2|f_i(y) - \mu_i h^*(y)|^2 + 2\nu c_3^2 9^{k+2} \mu_i^2 |y|^4.$$

Heading into the home stretch, we have

(57) $$\int_{\underline{\underline{B}}(0,9\theta)} \operatorname{dist}(x, A_i^*)^2 d\|V_i\| x$$
$$\leq \int_{\underline{\underline{B}}(0,9\theta) \cap X_i} \operatorname{dist}(x, A_i^*)^2 d\|V_i\| x$$
$$+ \int_{(C \sim X_i) \cap \underline{\underline{B}}(0,9\theta)} \operatorname{dist}(x, A_i^*)^2 d\|V_i\| x.$$

Using (28), (56), and (27),

(58) $$\int_{\underline{\underline{B}}(0,9\theta) \cap X_i} \operatorname{dist}(x, A_i^*)^2 d\|V_i\| x$$
$$\leq \int_{\underline{\underline{B}}(0,9\theta) \cap Y_i} \sum_m \operatorname{dist}(F_i(y)_m, A_i^*)^2 |\Lambda_k DF_i(y)_m| d\mathscr{H}^k y$$
$$\leq 2(c(\nu)+1)^k \Big[\int_D |f_i(y) - \mu_i h^*(y)|^2 d\mathscr{H}^k y$$
$$+ \nu c_3^2 9^{k+2} \mu_i^2 \int_{\underline{\underline{B}}^k(0,9\theta)} |y|^4 d\mathscr{H}^k y \Big].$$

By simple calculus,

(59) $$\int_{\underline{\underline{B}}^k(0,9\theta)} |y|^4 d\mathscr{H}^k y = (k\underline{\underline{\alpha}}/(k+4))(9\theta)^{k+4}.$$

Define

(60) $$Z_i = \{z \in \underline{\underline{B}}(0,9\theta) : |T^{\perp}(z)| > \mu^{1/(k+2)}\}$$

Then, using (54) and (60)

(61) $$\int_{(C\sim X_i)\cap\underline{\underline{B}}(0,9\theta)} \operatorname{dist}(x,A_i^*)^2 d\|V_i\|x$$

$$\leq \int_{(C\sim X_i)\cap\underline{\underline{B}}(0,9\theta)} 2|T^{\perp}(x)|^2 + 8c_3^2 7^{k+2}\mu_i^2 d\|V_i\|x$$

$$\leq 2\int_{Z_i} |T^{\perp}(x)|^2 d\|V_i\|x$$

$$+ (2\mu_i^{2/(k+2)} + 8c_3^2 9^{k+2}\mu_i^2)\|V_i\|(C\sim X_i).$$

We shall now show that for sufficiently large i

(62) $$\|V_i\|Z_i \leq (9\theta)^{k+4}\mu_i^2.$$

Suppose not. Then it follows from the Besicovitch covering theorem 2.2 and (15) that there is $z \in Z_i$ with $\Theta^k(\|V_i\|,z) \geq 1$ such that for $0 < r < |T^{\perp}(z)|/2$,

(63) $$\|\delta V_i\|\underline{\underline{B}}(z,r) \leq \underline{\underline{B}}(n)\|\delta V_i\|\underline{\underline{B}}(a_i,9)\|V_i\|\underline{\underline{B}}(z,r)/\|V_i\|Z_i$$

$$\leq \underline{\underline{B}}(n)M\mu_i^2 9^{k-1}\mu_i^{-2}(9\theta)^{-k-4}\|V_i\|\underline{\underline{B}}(z,r).$$

Letting γ correspond to $\lambda = 1/2$ and $\nu = 1$ in the monotonicity lemma 5.3, we have, for sufficiently large i,

$$(64) \qquad r\|\delta V_i\|\underline{\underline{B}}(z,r) \leq (\mu_i^{1/(k+2)}/2), \|\delta V_i\|\underline{\underline{B}}(z,r) < \gamma\|V\|\underline{\underline{B}}(z,r).$$

So by the monotonicity theorem 5.3,

$$(65) \qquad \|V_i\|\,\underline{\underline{B}}(z,|T^{\perp}(z)|/2) \geq (\underline{\underline{\alpha}}/2)|T^{\perp}(z)|^k 2^{-k}.$$

Recalling the definition of μ_i^2 from (14) and the definition of z_i from (60), we have

$$(66) \qquad 9^{k+2}\mu_i^2 > (\underline{\underline{\alpha}}/2)|T^{\perp}(z)|^{k+2}2^{-k-2}$$

$$> (\underline{\underline{\alpha}}/2)\mu_i 2^{-k-2},$$

which contradicts $\lim_{i\to\infty} \mu_i = 0$. Therefore (62) holds.

Combining (57), (58), (35), (59), (61), (26), (62), (9), and (14) yields

$$(67) \qquad \limsup_{i\to\infty} \mu_i^{-2}(9\theta)^{-k-2} \int_{\underline{\underline{B}}(0,7\theta)} \operatorname{dist}(x,A_i^*)^2 d\|V_i\|x$$

$$\leq [2+2(c(\nu)+1)^k \nu \underline{\underline{c}}_3^2 9^{k+2} k\underline{\underline{\alpha}}/(k+4)](9\theta)^{k+4}(9\theta)^{-k-2},$$

which, together with (36) and (48), contradicts (8) and (16).

□

<u>Remark</u>: The theorem remains true if hypothesis (6) is replaced by

(6') $$R^{-k+2} \int_{\underline{\underline{B}}(a,R)} |\underline{\underline{h}}(V,x)|^2 d\|V\|x < \eta\mu^2.$$

5.7. Flatness.

Theorem: If $V \in \underline{\underline{IV}}_k(\underline{\underline{R}}^n)$ and $\|\delta V\|$ is a Radon measure, then for V almost all $(y,T) \in \underline{\underline{G}}_k(\underline{\underline{R}}^n)$

(1) $$\lim_{r\to 0} r^{-k-3} \int_{\underline{\underline{B}}(y,r)} |T^{\perp}(x-y)|^2 d\|V\|x = 0.$$

Proof: For V almost all (y,T) we know that

(2) $$\Theta^k(\|V\|,y) \in \underline{\underline{N}},$$

(3) $$\mathrm{Tan}^k(\|V\|,y) = T,$$

(4) $$\underline{\underline{h}}(V,y) \in \underline{\underline{R}}^n,$$

and hence for V almost all (y,T),

(5) $$\lim_{r\to 0} r^{-k}\|V\|\{\alpha \in \underline{\underline{B}}(y,r): \Theta^k(\|V\|,y) \neq \Theta^k(\|V\|,x)\} = 0,$$

(6) $$\lim_{r\to 0} r^{-k-2} \int_{\underline{\underline{B}}(y,r)} |T^{\perp}(x-y)|^2 d\|V\|x = 0, \text{ and}$$

(7) $$\sup_{0<r<1} \{r^{-k}\|\delta V\|\underline{\underline{B}}(y,r)\} = B \text{ for some } B < \infty$$

Suppose y and T satisfy (2)-(7). Assume $y = 0$, let

$\nu = \Theta^k(\|V\|,0)$, define

(8) $$\mu(r)^2 = \inf_{A \in \underline{\underline{A}}(n,k)} r^{-k-2} \int_{\underline{\underline{B}}(y,r)} \operatorname{dist}(x,A)^2 d\|V\|x,$$

and let the infimum be obtained for $A(r)$. By the existence of a tangent plane at y,

(9) $$\lim_{r\to 0} v^{-1}\operatorname{dist}(0,A(r)) = 0 \quad \text{and}$$

(10) $$\lim_{r\to 0} r^{-2}\mu(r)^2 = 0.$$

We wish to show that $\lim_{r\to 0} r^{-1}\mu(r)^2 = 0$.

If $\limsup_{r\to 0} r^{-1}\mu(r)^2 > 0$, then there is $\varepsilon > 0$ such that for arbitrarily small v

(11) $$r^{-1}\mu(r)^2 > \varepsilon.$$

Choose $\theta < c_4^{-4}$ and let $M = \theta^{-k-3}\varepsilon^{-1}B$. Let η be as found in 5.6, and choose $R_0 > 0$ so that 5.6 (2), (3), (4), and (5) hold for $a = 0$, $0 < r \leq R_0$, and $A = A(r)$. Choose $m \in \underline{\underline{N}}$ so that

(12) $$\theta^{1-m/2}\varepsilon R_0 > \eta.$$

Suppose $0 < r < \theta^m R_0$ and $r^{-1}\mu(r)^2 > \varepsilon$. By the minimality of $A(r)$

(13) $$r^{-k-2}\mu(r)^2 < (r/\theta)^{-k-2}\mu(r/\theta)^2.$$

Hence

(14) $$(r/\theta)^{-1}\mu(r/\theta)^2 > \theta^{k+\varepsilon}\varepsilon.$$

We carefully chose M so that

(15) $$(r/\theta)^{-k+1}\|\delta V\|\underline{\underline{B}}(0,r/\theta) < M\mu(r/\theta)^2,$$

which is hypothesis (6) of 5.6. Hence 5.6 says that

(16) $$\mu(r)^2 \leq c_4^2\theta^2\mu(r/\theta)^2,$$

or, by choice of θ and $\mu(r)^2$,

(17) $$(r/\theta)^{-1}\mu(r/\theta)^2 > \theta^{-1/2}\varepsilon.$$

Thus we may repeat this process until we get $p \in N$ with $p \geq m$, $\theta R_0 \leq r/\theta^p \leq R_0$, and

(18) $$(r/\theta^p)^{-1}\mu(r/\theta^p)^2 > \theta^{-p/2}\varepsilon.$$

But by the choice of m, we then have

$$\mu(r/\theta^p)^2 > \theta R_0\theta^{-m/2}\varepsilon > \eta,$$

which contradicts $\mu(R)^2 < \eta$ for $R < R_0$. Hence $\lim_{r\to 0} r^{-1}\mu(r)^2 = 0$.

It remains to replace $A(r)$ by T. We do this by comparing $A(r)$ and $A(r/2)$. For each $r > 0$, let $A(r) = T(r) + b(r)$, where $T(r) \in \underline{G}(n,k)$ and $b(r) \in T(r)^{\perp}$. For small r, where V is nearly a k-plane, a little geometry shows that for some constant c we must have

$$|b(r)-b(r/2)| < cv\mu(r) \quad \text{and}$$

$$\|T(r)-T(r/2)\| \leq c\mu(r).$$

Because there is a tangent plane at 0, clearly

$$\lim_{r\to 0} b(r) = 0, \quad \lim_{r\to 0} T(r) = T.$$

Thus, when $\mu(s)^2 < s$ for all $s < r$

$$|b(r)| \leq \sum_{m=0}^{\infty} cr2^{-m}\mu(r/2^m)$$

$$\leq \sum_{m=0}^{\infty} cr^{3/2}2^{-3m/2} \leq 2cr^{3/2} \quad \text{and}$$

$$\|T(r)-T\| \leq \sum_{m=0}^{\infty} c\mu(r/2^m)$$

$$\leq \sum_{m=0}^{\infty} cr^{1/2}2^{-m/2} \leq 4cr^{1/2}.$$

Hence $A(r)$ is close enough to T that $r^{-1}\mu(r)^2 \to 0$ implies

$$\lim_{r\to 0} r^{-k-3} \int_{\underline{B}(0,r)} |T^{\perp}(x)|^2 d\|V\|x = 0. \qquad \square$$

Remark: When $\|\delta V\|_{sing} = 0$ and $\underline{h}(V,\cdot)$ is locally square integrable with respect to $\|V\|$, then r^{-k-3} may be replaced by $r^{-k-4+\delta}$ for any $\delta > 0$.

5.8. Perpendicularity.

Theorem: *If* $V \in \mathbf{IV}_k(\mathbf{R}^n)$ *and* $\|\delta V\|$ *is a Radon measure, then*

(1) $$T(\underline{h}(V,y)) = 0$$

for V *almost all* $(y,T) \in \mathbf{G}_k(\mathbf{R}^n)$.

Proof: To prove (1), it is sufficient to show that

(2) $$\lim_{r\to 0} r^{-k}\delta V(\chi^2(r,\cdot-y)w) = 0$$

for every $w \in T$ with $|w| = 1$.

From chapter 2 and 5.7 we know that V almost all $(y,T) \in \mathbf{G}_k(\mathbf{R}^n)$ satisfy

(3) $$\underline{h}(V,y) \in \mathbf{R}^n,$$

(4) $$\Theta^k(\|V\|,y) \in \mathbf{N},$$

(5) $$\operatorname{Tan}^k(\|V\|,y) = T;$$

(6) $$\sup_{0<r<1} \{r^{-k}\|\delta V\|\mathbf{B}(y,r)\} = B \text{ for some } B < \infty, \text{ and}$$

(7) $$\lim_{r\to 0} r^{-k-3} \int_{\underline{\underline{B}}(4,r)} |T^{\perp}(x)|^2 d\|V\|x = 0.$$

Suppose (y,T) satisfies (3)-(6), $w \in T$, $|w| = 1$, and $\varepsilon > 0$. We may assume that $y = 0$ and $T = \underline{\underline{e}}_1 \wedge \ldots \wedge \underline{\underline{e}}_k$. Let $\nu = \Theta^k(\|V\|,0)$. For sufficiently small $r > 0$ we have:

(8) $$(\nu - 1/4)\underline{\underline{\alpha}} r^k < \|V\|\underline{\underline{B}}(0,r),$$

$$\|V\|\underline{\underline{B}}(0,3r) \leq (\nu + 1/4)\underline{\underline{\alpha}}(3r)^k, \text{ and}$$

(9) $$\int_{\underline{\underline{B}}(0,9r)} |T^{\perp}(x)|^2 d\|V\|x < \varepsilon r^{k+3}.$$

Let $f\colon T \to \underline{\underline{M}}_\nu$ and $F\colon T \to T \times M_\nu$ be the Lipschitz approximations constructed in 5.4, scaled down to $\underline{\underline{B}}(0,r)$, and let X and Y also be as in 5.4.

Let $\zeta(x) = \chi(r,x)^2$. From 2.9 we have

(10) $$\delta V(\zeta w) = \int D\zeta(x) \otimes w \cdot S \, dV(x,S),$$

and by symmetry we have

(11) $$\nu \int_T D\zeta(y) \cdot w \, d\mathscr{H}^k y = 0.$$

To connect (10) and (11) we define quantities $a_1, \ldots, a_5$ as follows:

(12) $$a_1 = \int D\zeta(x)\otimes w \cdot (S-T)\,dV(x,S),$$

$$a_2 = \int_{\underline{\underline{B}}(0,r)\sim X} D\zeta(x)\cdot w \; d\|V\|x,$$

$$a_3 = \int_X D\zeta(x)\cdot w(1-|\Lambda_k T\cdot S|^{-1})\,dV(x,S),$$

$$a_4 = \int_Y \sum_m [D\zeta(F(y)_m)-D\zeta(y)]\cdot w \; d\mathscr{H}^k y, \quad \text{and}$$

$$a_5 = \int_{\underline{\underline{B}}^k(0,r)\sim Y} \nu D\zeta(y)\cdot w \; d\mathscr{H}^k y.$$

By (10), (11), and $D\zeta(x)\otimes w \cdot T = D\zeta(x)\cdot w$,

(13) $$\delta V(\zeta w) = a_1 + \cdots + a_5.$$

Using the tilt lemma 5.4 with $\phi(x) = \chi(8r,x)$, $p = 1$, $\alpha = B(8r)^k$ from (6), $\mu^2 = \varepsilon r^{k+3}$ from (9), $\xi^2 = (p/8r)^2\varepsilon r^{k+3}$ from (9) and the properties of χ, we get

(14) $$\int_{\underline{\underline{B}}(0,7r)} \|S-T\|^2 dV(x,S)$$

$$\leq (B8^k r^k)^{2/3}(\varepsilon r^{k+3})^{1/3} + (\rho^2\varepsilon/4)r^{k+1}$$

$$\leq (B^{2/3}4^k\varepsilon^{1/3} + \rho^2\varepsilon/4)r^{k+1}.$$

We estimate: using $|S(w)-w| \leq \|S-T\|^2$ and (14),

(15) $$|a_1| \leq \int_{\underline{\underline{B}}(0,r)} (2\rho/r)\|S-\tilde{T}\|^2 dV(x,S)$$

$$\leq 2\rho(B^{2/3}4^k\varepsilon^{1/3} + \rho^2\varepsilon/4)r^k;$$

using 5.4 (10) with appropriate scaling, (6), (9), and (14),

(16) $$|a_2| + |a_5| \le \nu(2\rho/r)P[(B7^k r)^{k/(k-1)}$$

$$+ (B^{2/3}4^k\varepsilon^{1/3} + \rho^2\varepsilon/4)r + (\rho/8)^2\varepsilon r]r^k;$$

using $|1-|\Lambda_k\tilde{T}\circ S|^{-1}| < k\|S-T\|^2$ and (14),

(17) $$|a_3| \le (2\rho/r)(B^{2/3}4^k\varepsilon^{1/3} + \rho^2\varepsilon/4)r^{k+1};$$

and using the properties of χ, (9), and $F(y)_m\cdot w = y\cdot w$,

(18) $$|a_4| = \left|\int_Y \sum_m [|D\zeta(y)|y/|y| - |D\zeta(F(y)_m)|F(y)_m/|F(y)_m|]\cdot w \, d\mathscr{H}^k y\right|$$

$$\le \int_Y \sum_m |(|D\zeta(y)| - |D\zeta(F(y)_m)|)y/|y|$$

$$+ |D\zeta(F(y)_m)|y(|y|^{-1} - |F(y)_m|^{-1})| d\mathscr{H}^k y$$

$$\le \int_Y \sum_m \sup_{x\in\mathbf{R}^n} \{D^2\zeta(x)\} ||y| - |F(y)_m||$$

$$+ \sup_{x\in\mathbf{R}^n} \{D\zeta(x)/|x|\} ||y| - |F(y)_m|| d\mathscr{H}^k y$$

$$\le \int_{\mathbf{B}(0,r)} (4\rho/r^2)|T^\perp(x)|^2 + (2\rho/r^2)|T^\perp(x)|^2 d\|V\|x$$

$$\le 6\rho\varepsilon r^{k+1}.$$

The estimates (15)-(18) are all of order no more than $\varepsilon^{1/3}r^k$, so as $\varepsilon \to 0$,

$$\lim_{r\to 0} r^{-k}\delta V(\chi^2(r,\cdot)w) = 0. \qquad \square$$

6. Regularity

In this chapter we investigate the regularity of integral varifolds moving by their mean curvature. Because of the close relationship to parabolic partial differential equations, in particular the heat equation, one would expect that such a varifold would be an infinitely differentiable manifold, except perhaps on a set of $\mathscr{H}^k$ measure zero where several sheets join.

We shall prove in 6.13 that an integral varifold moving by its mean curvature has the desired regularity, but only under the hypothesis that the varifold has unit density almost everywhere at almost all times. Indeed, it is not even known if a stationary integral varifold, i.e. one with zero first variation, is regular when multiple densities are permitted. The next section describes an example showing the problems stemming from multiple densities. The existence theory of chapter 4 is not yet known to produce varifolds with unit density, as remarked in 4.9. However, the physical surfaces that varifolds model always seem to have unit density.

The idea of the proof is to show that a flat enough piece of a varifold moving by its mean curvature can be represented as the graph of a Lipschitz function; the theory of parabolic partial differential equations then quickly gives the infinite differentiability. To get the Lipschitz representation we show that surplus mass quickly disappears, that a mass deficit means holes which cause the varifold to pop like a soap film, and that otherwise things tend to average out, analagous to the diffusion of heat.

6.1. A multiple density example.

We will construct an integral varifold $V \in \underline{\underline{IV}}_2(\underline{\underline{R}}^3)$ with bounded mean curvature such that there is a set $A \subset \underline{\underline{R}}^n$ with $\|V\|A > 0$ for which no element x of A has a neighborhood in which V can be represented by the graph of a function, even a multiple valued function.

It is well known that a catenoid has zero mean curvature. Having in mind a radius $R > 0$ and an upper bound B for the mean curvature, one can take a catenoid with a very small central hole and gradually bend the two sheets together away from the hole to get a varifold that has mean curvature bounded by B, that is a double density plane outside radius R, and has a hole in the middle.

To construct V, start with a double density plane in $\underline{\underline{R}}^3$. Remove a disjoint collection of disks whose union is dense in the plane yet leaves behind a set A of positive area. Replace each disk with a section of bent catenoid with a hole so that the edges match smoothly. The resulting integral varifold V has integral densities and bounded mean curvature, yet if $x \in A$, then V has holes in every neighborhood of A and hence cannot be represented as the graph of a function.

This example is not a varifold moving by its mean curvature (the construction does not work for zero mean curvature), and it would mostly instantaneously vanish under the construction given in chapter 4, but it cannot yet be ruled out that some slowly changing version of V would be moving by its mean curvature. In any case, V does show the need for the unit density hypothesis for the method used in this chapter. Note that in V the holes are surrounded by single density, and the bad points of A are double density. By eliminating the double density, we eliminate the possibility of holes with small mean curvature.

6.2. Regularity and square integrable mean curvature.

It was shown in [AW1 chap. 8] under the unit density hypothesis that a k-dimensional varifold is almost everywhere a Holder continuously differentiable manifold if the mean curvature is locally integrable to a power greater than k. We know from 3.4 that a varifold moving by its mean curvature has locally square integrable mean curvature. Hence the above result gives some degree of regularity only for $k = 1$.

To see what happens in higher dimensions, note that a k-sphere of radius R has mean curvature k/R and hence total squared mean curvature of $k(k+1)\underline{\underline{\alpha}}R^{k-2}$. Thus for $k > 2$ one can scatter infinitely many tiny spheres densely throughout space while keeping the mean curvature square integrable. The support of the varifold would be the whole space, so there would be no chance of regularity.

We can prove a regularity theorem for a varifold moving by its mean curvature because tiny spheres and like things quickly wipe themselves out.

6.3. *Clearing out.*

This lemma shows that a region with little mass in it quickly becomes empty.

Lemma: *If* $2 < m < \infty$ *then there exists* $c(m) > 0$ *such that if* V_t *is an integral varifold moving by its mean curvature,*

(1) $$0 \leq \eta < \infty, \quad t_0 \in \mathbb{R}^+, \quad 0 < R_0 < \infty, \quad a \in \mathbb{R}^n,$$

(2) $$\phi(x) = \begin{cases} 1-|x-\alpha|^2/R_0^2 & \text{\underline{for}} \quad 0 \leq |\alpha-a| \leq R_0 \\ 0 & \text{\underline{for}} \quad |x-a| \geq R_0, \end{cases}$$

(3) $$\|V_{t_0}\|(\phi^m) < \eta R_0^k,$$

(4) $$t_1 > t_0 + c(m)\eta^{2/(k+2m)}R_0^2 \quad \text{\underline{and}}$$

$$R_1^2 = R_0^2 - 4k(t_1-t_0),$$

then

(5) $$\|V_{t_1}\|\mathbb{B}(a,R_1) = 0.$$

Proof: We will use a fast shrinking test function, similar to the barrier function of 3.7. Mass near the edge will be left behind, and mass in the interior must have high mean curvature, which will wipe it out.

Because of the behavior of the various quantities under translation and homothety, it suffices to prove the lemma with $t_0 = 0$, $R_0 = 1$, and $a = 0$. We shall let $\varepsilon = 1/(k+2m)$.

For $0 < t < 1/4k$, define $R(t)$ and $\phi(t,\cdot)\colon \underline{\underline{R}}^n \to R^+$ as follows:

$$R(t)^2 = 1-4kt \tag{2}$$

$$\phi(t,x) = \begin{cases} R(t)^2 - |x|^2 & \text{if } 0 \le |x| \le R(t) \\ 0 & \text{if } |x| \ge R(t). \end{cases}$$

Also let $\xi(t) = \|V_t\|(\phi(t,\cdot)^m)$. In what follows, ϕ always refers to the time varying function just defined, and ξ means $\xi(t)$.

By 3.5,

$$\overline{D}\xi(t) \le \delta(V_t,\phi^m)(\underline{\underline{h}}(V_t,\cdot)) + \|V_t\|(\partial\phi^m/\partial t), \tag{3}$$

so by 3.2(3) and the perpendicularity of mean curvature,

$$\overline{D}\xi(t) \le -\int |\underline{\underline{h}}(V_t,x)|^2\phi^n(t,x)\,d\|V_t\|x \tag{4}$$

$$+ \int \underline{\underline{h}}(V_t,x)\cdot D\phi^m(t,x)\,d\|V_t\|x + \|V_t\|(\partial\phi^m/\partial t).$$

By the definition of mean curvature in 2.9,

(5) $$\int \underline{\underline{h}}(V_t,x)\cdot D\phi^m(t,x)\,d\|V_t\|x = -\delta V_t(D\phi^m)$$

$$= -\int D^2\phi^m(t,x)\cdot S\ dV_t(x,S).$$

$$= -\int [4m(m-1)\,\phi^{m-2}(t,x)\,(x\otimes x) - 2m\phi^{m-1}(t,x)\underline{\underline{I}}]\cdot S\ dV_t(x,S)$$

$$= \int -4m(m-1)\,\phi^{m-2}(t,x)\,|S(x)|^2 - 2km\phi^{m-1}(t,x)\,dV(x,S).$$

Since $\partial\phi^m/\partial t = -4km\phi^{m-1}$, we have

(6) $$\overline{D}\xi(t) \le -\int |\underline{\underline{h}}(V_t,x)|^2\phi^m(t,x)\,d\|V_t\|x$$

$$-4m(m-1)\int |S(x)|^2\phi^{m-2}(t,x)\,dV_t(x,S)$$

$$-2km\int \phi^{m-1}(t,x)\,d\|V_t\|x.$$

We wish to show $\overline{D}\xi(t) \le -c\xi(t)^{1-2\varepsilon}$ for a yet to be determined constant c. If not, then we will have, for some t,

(7) $$\int |\underline{\underline{h}}(V_t,x)|^2\phi^m(t,x)\,d\|V_t\|x < c\xi(t)^{1-2\varepsilon},$$

$$\int |S(x)|^2\phi^{m-2}(t,x)\,dV_t(x,S) < c\xi(t)^{1-2\varepsilon}, \quad \text{and}$$

$$\int \phi^{m-1}(t,x)\,d\|V_t\|x < c\xi(t)^{1-2\varepsilon}.$$

It follows from the Besicovitch covering theorem 2.2 that there is a point $b \in R^n$ such that $\Theta^k(\|V_t\|,b) \ge 1$ and

$$(8)\qquad \int_{\mathbf{B}(b,r)} |\mathbf{h}(V_t,x)|^2\phi^m(t,x)\,d\|V_t\|x < 3c\mathbf{B}(n)\xi^{-2\varepsilon}\int_{\mathbf{B}(b,r)} \phi^m d\|V_t\|,$$

$$(9)\qquad \int_{\mathbf{B}(b,r)} |S(x)|^2\phi^{m-2}(t,x)\,dV_t(x,S) < 3c\mathbf{B}(n)\xi^{-2\varepsilon}\int_{\mathbf{B}(b,r)} \phi^m d\|V_t\|,$$

$$(10)\qquad \int_{\mathbf{B}(b,r)} \phi^{m-1}(t,x)\,d\|V_t\|x < 3c\mathbf{B}(n)\xi^{-2\varepsilon}\int_{\mathbf{B}(b,r)} \phi^m d\|V_t\|.$$

for every $0 < r < \infty$. From (10), as $r \to 0$, we see that $\phi(t,b) > \xi^{2\varepsilon}/3c\mathbf{B}(n)$.

Now consider $V_t \llcorner \phi^m$. If g is a test vectorfield, then

$$(11)\qquad \delta(V_t \llcorner \phi^m)(g) = \int Dg(x)\cdot S\phi^m(t,x)\,dV_t(x,S)$$

$$= \int D(\phi^m(t,x)g(x))\cdot S - g(x)\otimes D\phi^m(t,x)\cdot S\ dV_t(x,S)$$

$$= \int -\mathbf{h}(V_t,x)\cdot g(x)\phi^m(t,x) - g(x)\cdot S(D\phi^m(t,x))\,dV_t(x,S).$$

Therefore, since $D\phi^m(t,x) = m\phi^{m-1}(t,x)(-2x)$,

$$(12)\qquad \|\delta(V_t \llcorner \phi^m)\|\mathbf{B}(b,r) \le \int_{\mathbf{B}(b,r)} |\mathbf{h}(V_t,x)|\phi^m(t,x)\,d\|V_t\|x$$

$$+ 2m\int_{\mathbf{B}(b,r)} |S(x)|\phi^{m-1}(t,x)\,dV(x,S).$$

for all $r > 0$. Then by Schwartz' inequality, (8), and (9),

$$(13)\qquad \|\delta(V_t \llcorner \phi^m)\|\underline{\underline{B}}(b,r)$$

$$\leq \left[\int_{\underline{\underline{B}}(b,r)} |\underline{\underline{h}}(V_t,x)|^2\phi^m(t,x)\,d\|V_t\|x \int_{\underline{\underline{B}}(b,r)} \phi^m(t,x)\,d\|V_t\|x\right]^{1/2}$$

$$+\ 2m\left[\int_{\underline{\underline{B}}(b,r)} |S(x)|^2\phi^{m-2}(t,x)\,dV(x,S) \int_{\underline{\underline{B}}(b,r)} \phi^m(t,x)\,d\|V_t\|x\right]^{1/2}$$

$$\leq [3c\underline{\underline{B}}(n)]^{-1/2}\xi^{-\varepsilon}\|V_t \llcorner \phi^m\|\underline{\underline{B}}(b,r)$$

$$+\ 2m[3c\underline{\underline{B}}(n)]^{1/2}\xi^{-\varepsilon}\|V_t \llcorner \phi^m\|\underline{\underline{B}}(b,r)$$

for all $r > 0$. Let $r(b) = [3c\underline{\underline{B}}(n)]^{-1/2}\xi^{\varepsilon}/(2m+1)$. Then the monotonicity lemma 4.17 and (13) imply

$$(14)\qquad \|V_t \llcorner \phi^m\|\underline{\underline{B}}(b,r(b)) \geq e^{-1}\underline{\underline{\alpha}}r(b)^k\Theta^k(\|V_t \llcorner \phi^m\|,b).$$

By the density hypothesis in b, we clearly have

$$(15)\qquad \Theta^k(\|V_t \llcorner \phi^m\|,b) \geq \phi^m(t,b) \geq [\xi^{2\varepsilon}/3c\underline{\underline{B}}(n)]^m.$$

Hence (14) yields

$$(16)\qquad \|V_t \llcorner \phi^m\|\underline{\underline{B}}(b,r(b)) \geq e^{-1}\underline{\underline{\alpha}}[3c\underline{\underline{B}}(n)]^{-m-k/2}(2m+1)^{-k}\xi^{(m+k/2)\varepsilon}$$

$$\geq e^{-1}\underline{\underline{\alpha}}[3c\underline{\underline{B}}(n)]^{-m-k/2}(2m+1)^{-k}\xi.$$

Thus, if c is small enough, we have a contradiction to $\xi = \|V_t \llcorner \phi^m\|\underline{\underline{R}}^n$.

Now that we have established

(17) $$D\xi(t) \le -c\xi(t)^{1-2\varepsilon}$$

we need only integrate to find that $\xi(t) = 0$ for $t \ge \xi(0)^{2\varepsilon}/2c\varepsilon$. Since $\xi(0) \le \eta$, we have established the lemma with $c(m) = 1/2c\varepsilon$. □

6.4. Cylindrical growth rates.

We will later be dealing with nearly flat varifolds and cylinders nearly perpendicular to them, and we will need estimates of the rate of growth of the mass in a cylinder as a function of radius. Note that we get both upper and lower bounds, as contrasted with the lower bounds of the spherical monotonicity theorem 4.17.

Theorem: Suppose

(1) $T \in \mathbf{G}(n,k)$, $0 < R_1 < R_2 < \infty$, $0 \le \alpha < \infty$, $0 \le \beta < \infty$,

(2) $V \in \mathbf{IV}_k(\mathbf{R}^n)$ and $\operatorname{spt}\|V\| \cap \mathbf{C}(T,0,R_2)$ is compact,

(3) $\phi \in \mathbf{C}_0^3(\mathbf{R}^n, R^+)$, $\operatorname{spt}\phi < \mathbf{C}(T,0,1)$,

$\phi(x)$ depends only on $|T(x)|$,

(4) $\int |\mathbf{h}(V,x)|^2 \phi(x/r)\, d\|V\|x < \alpha^2 r^k$ for $R_1 \le r \le R_2$, and

(5) $\int \|S-T\|^2 \phi(x/r)\, dV(x,S) < \beta^2 r^k$ for $R_1 \le r \le R_2$.

Then

$$(6) \qquad |R_2^{-k}\|V\|(\phi(x/R_2)) - R_1^{-k}\|V\|(\phi(x/R_1))|$$

$$< k\beta^2\log(R_2/R_1) + \alpha\beta(R_2-R_1) + \beta^2.$$

Proof: We assume that $a = 0$. Recall that under our assumptions, (4) implies $\|\delta V\|_{sing}\underline{\underline{C}}(T,0,R_2) = 0$.

For $R_1 \le r_1 \le R_2$, 2.9 and the boundedness of $\operatorname{spt}\|V\| \cap \underline{\underline{C}}(T,0,R_2)$ guarantee the validity of

$$(7) \qquad \delta V(\phi(x/r)T(x)/r) = r^{-1}\int T\cdot S\phi(x/r)\,dV(x,S)$$

$$+ r^{-1}\int T(x) \otimes D\phi(x/r)\cdot S\,dV(x,S).$$

Because $\phi(x)$ depends only on $|T(x)|$, we have

$$(8) \qquad D\phi(x/R) = r(\partial\phi(x/r)/\partial r)T(x).$$

Using the perpendicularity of mean curvature, Schwarz' inequality, (4), and (5),

$$(9) \qquad |\delta V(\phi(x/r)T(x)/r)|$$

$$= |\int \underline{\underline{h}}(V,x)\cdot S^{\perp}(T(x)/r)\phi(x/r)\,dV(x,S)|$$

$$\le [\int |\underline{\underline{h}}(V,x)|^2\phi(x,r)\,d\|V\|x \cdot \int \|S-T\|^2\phi(x/r)\,dV(x,S)]^{1/2}$$

$$\le \alpha\beta r^k.$$

It is not too hard to see that

$$(10) \qquad |T\cdot S-k|S(T(x))|^2|T(x)|^{-2}| \le k\|S-T\|^2.$$

We use (8), (9), and (10) in (7) and rearrange:

$$(11) \qquad |(d/dr)\int |S(T(x))|^2|T(x)|^{-2}\phi(x/r)\,dV(x,S)$$

$$- (k/r)\int |S(T(x))|^2|T(x)|^{-2}\phi(x/r)\,dV(x,S)|$$

$$\le r^{-1}k\beta^2 r^k + \alpha\beta r^k.$$

Integrating this from R_1 to R_2 yields

$$(12) \qquad |r^{-k}\int |S(T(x))|^2|T(x)|^{-2}\phi(x/r)\,dV(x,S)|_{R_1}^{R_2}|$$

$$\le k\beta^2\log(R_2/R_1) + \alpha\beta(R_2-R_1).$$

Because

$$1-|S(T(x))|^2|T(x)|^{-2} \le \|S-T\|^2$$

we can combine (5) with (12) to get

$$|R_2^{-k}\|V\|(\phi(x/R_2))-R_1^{-k}\|V\|(\phi(x/R_1))|$$

$$\le k\beta^2\log(R_2/R_1)+\alpha\beta(R_2-R_1)+\beta^2. \qquad \square$$

6.5. Expanding holes.

Here we show that, for a nearly flat varifold, thin spots will expand and thick spots will contract. For later convenience, the spots expand or contract non-isotropically.

Lemma: *If* V *is an integral varifold moving by its mean curvature,* $0 \le \mu < \infty$, $T \in \underline{\underline{G}}(n,k)$, $a \in \underline{\underline{R}}^n$, $0 \le t_1 < t_2 < \infty$, $0 < R_1$, $R_2 < 1$, $\sigma = (R_2^2 - R_1^2)/(t_2 - t_1)$, $R(t)^2 = R_1^2 + \sigma(t - t_1)$, $\phi_t(x) = \chi_T([x/R(t)] - a)$,

(1) $$\int_{\underline{\underline{C}}(T,0,R(t))} |T^{\perp}(x)|^2 d\|V_t\|x = \mu^2(t) \le \mu^2 R(t)^{k+2} \quad \text{and}$$

(2) $$\alpha(t)^2 = \int |\underline{\underline{h}}(V_t,x)|^2 \phi_t^2(x)\, d\|V_t\|x$$

for $t \in [t_1, t_2]$, *then*

(3) $$\delta(V,\phi_t^2)(h(V_t,\cdot)) < -\alpha(t)^2/2 + 32\rho^2 R(t)^{-4}\mu(t)^2$$

for all $t \in [t_2, t_2]$ *and there is* $M < \infty$ *depending on* σ *such that*

(4) $$R_2^{-k}\|V_{t_2}\|(\phi_{t_2}^2) \le R_1^{-k}\|V_{t_1}\|(\phi_{t_1}^2) + M\mu^2|\log R_2/R_1|.$$

Proof: By 3.3 and because $D\phi_t$ lies in T, we have, using Minkowski's inequality,

(5) $$\delta(V_t,\phi_t^2)(\underline{\underline{h}}(V,\cdot)) = -\int |\underline{\underline{h}}(V_t,x)|^2\phi_t^2(x)\,d\|V_t\|x$$

$$+ \int 2\underline{\underline{h}}(V_t,x)S^{\perp}(D\phi_t(x))\phi_t(x)\,dV_t(x,S)$$

$$\leq -\alpha(t)^2+2\int |\underline{\underline{h}}(V_t,x)|\,\|S-T\|\,|D\phi_t(x)|\phi_t(x)\,dV(x,S)$$

$$\leq -\alpha(t)^2+(1/4)\int |\underline{\underline{h}}(V_t,x)|^2\phi_t^2(x)\,d\|V_t\|x$$

$$+ 4\int \|S-T\|^2|D\phi_t(x)|^2 dV(x,S).$$

A slight modification of the derivation of 5.5 yields

(6) $$\int |D\phi_t(x)|^2\|S-T\|^2 dV_t(x,S)$$

$$\leq 16\int |T^{\perp}(x)|^2|D|D\phi_t(x)||^2 d\|V_t\|x$$

$$+ 2[\int |\underline{\underline{h}}(V_t,x)|^2\phi_t^2(x)\,d\|V_t\|x\int |T^{\perp}(x)|^2|D\phi_t(x)|^4\phi_t^{-2}(x)\,d\|V\|x]^{1/2}$$

$$\leq 16\rho^2R(t)^{-4}\mu(t)^2 + 2\alpha(t)\rho R(t)^{-2}\mu(t)$$

$$\leq \alpha(t)^2/4 + 32\rho^2R(t)^{-4}\mu(t)^2.$$

Combining (5) and (6) yields (3).

For (4), we need to find $\|V_t\|(\partial\phi_t^2/\partial t)$. From the definition of ϕ_t, we get

(7) $$\partial\phi_t^2(x)/\partial t = -2\phi_t(x)D\phi_t(x)\cdot x(R'(t)/R(t))$$

We also have

$$(8)\qquad \delta V_t(\phi_t^2(x)T(x)) = -\int \underline{\underline{h}}(V_t,x)\cdot T(x)\,\phi_t^2(x)\,d\|V_t\|x$$

$$= \int 2\phi_t(x)D\phi_t(x) \otimes T(x)\cdot S + \phi_t^2(x)T\cdot S\; dV_t(x,S).$$

One finds from (8) that

$$(9)\qquad \int -2\phi_t(x)D\phi_t(x)\cdot x\; d\|V_t\|x \le k\|V_t\|(\phi_t^2)$$

$$+2\int \phi_t(x)D\phi_t(x)\cdot(S(T(x))-x)\,dV_t(x,S)$$

$$+\int \underline{\underline{h}}(V_t,x)\cdot T(x)\,\phi_t^2(x)\,d\|V_t\|x$$

$$+\int \phi_t^2(x)\,(T\cdot S-k)\,dV_t(x,S).$$

By the properties of $\underline{\underline{G}}(n,k)$ in 2.5, the perpendicularity of mean curvature, $\operatorname{spt}\phi_t \subset \underline{\underline{C}}(T,0,R(t))$, Minkowski's inequality, 5.

$$(10)\qquad \int -2\phi_t(x)D\phi_t(x)\cdot x\,d\|V_t\|x \le k\|V_t\|(\phi_t^2)$$

$$+\; 2\int \phi_t(x)\,|D\phi_t(x)|\,\|S-T\|^2\,|T(x)|\,dV_t(x,S)$$

$$+\int |\underline{\underline{h}}(V_t,x)|\,\|S-T\|\,|T(x)|\,\phi_t^2(x)\,d\|V_t\|x$$

$$+\int \phi_t^2(x)\,k\,\|S-T\|^2 dV_t(x,S)$$

$$\le k\|V_t\|(\phi_t^2)+(1+|\sigma|+k)\int \phi_t^2(x)\,\|S-T\|^2 dV_t(x,S)$$

$$+\; |R(t)^2/4\sigma|\int |\underline{\underline{h}}(V_t,x)|^2\phi_t^2(x)\,d\|V_t\|x$$

$$+ R(t)^2 \int |D\phi_t(x)|^2 \|S-T\|^2 dV_t(x,S)$$

$$\leq k\|V_t\|(\phi_t^2)+(k+|\sigma|+1)[2\alpha(t)\mu(t) + 16\rho^2R(t)^{-2}\mu(t)^2]$$

$$+ |R(t)^2\alpha(t)^2/4\sigma|+[2\alpha(t)\mu(t) + 16\rho^2R(t)^{-2}\mu(t)^2]$$

$$\leq k\|V_t\|(\phi_t^2)+|R(t)^2\alpha(t)^2/2\sigma|+(k+|\sigma|+2)^2|\sigma|R(t)^{-1}\mu(t)^2$$

$$+ (k+|\sigma|+2)16\rho^2R(t)^{-2}\mu(t)^2.$$

By 3.5, (3), (7), (10), and (1)

$$\text{(11)} \qquad \overline{D}\|V_t\|(\phi_t^2) \leq \delta(V_t,\phi_t^2)(\underline{\underline{h}}(V_t,x))+\|V_t\|(\partial\phi_t^2/\partial t)$$

$$\leq -\alpha(t)^2/2 + 32\rho^2\mu^2R(t)^{k-2}$$

$$+ k(R'(t)/R(t))\|V_t\|(\phi_t^2) + R'(t)R(t)\alpha(t)^2/2|\sigma|$$

$$+ (R'(t)/R(t))[(k+|\sigma|+2)^2|\sigma| + 16(k+|\sigma|+2)\rho^2]\mu^2.$$

Since $\sigma = 2R'(t)R(t)$, we have

$$\text{(12)} \qquad R(t)^{-k}\overline{D}\|V_t\|(\phi_t^2)-kR'(t)R(t)^{-k-1}\|V_t\|(\phi_t^2)$$

$$\leq |R'(t)R(t)^{-1}|[64\rho^2/|\sigma|+(k+|\sigma|+2)^2|\sigma| + 16(k+|\sigma|+2)\rho^2]\mu^2.$$

If we let M be the quantity in brackets, then integration of (12) gives

$$R(t_2)^{-k}\|V_{t_2}\|(\phi_{t_2}^2)-R(t_1)^{-k}\|V_{t_1}\|(\phi_{t_1}^2) \leq M\mu^2|\log R_2/R_1|.$$

□

6.6. Popping soap films.

The Lipschitz approximation theorem 5.4 shows that a nearly flat varifold either has nearly integral density ratios or else has considerable first variation. Here we show that in a moving varifold, this first variation quickly drives density ratios towards integers.

Lemma: *There are constants* $c_{10}, c_{11} < \infty$ *such that if* V_t *is an integral varifold moving by its mean curvature,* $\mathrm{spt}\|V_t\| \cap \underline{\underline{C}}(T,0,1)$ *is bounded,*

(1) $$T \in \underline{\underline{G}}(n,k), \quad 0 < t_1 < t_2 < \infty, \quad 0 \le \mu \le \infty, \quad \nu = 0,1, \text{ or } 2,$$

(2) $$\int_{\underline{\underline{C}}(T,0,1)} |T^{\perp}(x)|^2 d\|V_t\|x < \mu^2 \quad \text{for} \quad t_1 \le t \le t_2, \quad \text{and}$$

(3) $$\|V_{t_1}\|(\chi_T^2) \le (\nu+1)\underline{\underline{\beta}} - c_{10}\mu^2,$$

then for $t_1 + c_{11} < t < t_2$

(4) $$\|V_t\|(\chi_T^2) \le \nu\underline{\underline{\beta}} + c_{10}\mu^2.$$

Proof: Suppress the t variable temporarily. Define

(5) $$\alpha^2 = \int |\underline{\underline{h}}(V,x)|^2 \chi_T^2(x)\, d\|V\|x,$$

(6) $$\beta^2 = \int \|S-T\|^2 \chi_T^2(x)\, d\|V\|x.$$

Then the tilt lemma 5.5 and (2) say that

$$(7) \qquad \beta^2 \le 2\alpha\mu + 16\rho^2\mu^2.$$

Suppose that $\nu = 0, 1,$ or 2 and

$$(8) \qquad (\nu-1/2)\underline{\underline{\beta}} \le \|V\|(\chi_T^2) \le (\nu+1/2)\underline{\underline{\beta}}.$$

We want to apply the Lipschitz approximation theorem 5.4 to $\underline{\underline{B}}(0,1/9)$, so we must check 5.4(2). Define $r_1 = 1 + 1/100k$. Then, using a little geometry,

$$(9) \qquad \|V\|\underline{\underline{B}}(0,1/9) \ge \|V\|\underline{\underline{C}}(T,0,r_1/9)$$

$$- \|V\|\{x \in \underline{\underline{C}}(T,0,r_1/9): |T^{\perp}(x)| > (r_1-1)^{1/2}/9\}$$

$$\ge \|V\|(\chi_T^2(r_1/9,\cdot)) - 8100k\,\mu^2 \quad \text{and}$$

$$(10) \qquad \|V\|\underline{\underline{B}}(0,1/3) \le \|V\|(\chi_T^2(r_1/3,\cdot)).$$

By the cylindrical growth lemma 6.4 and (8),

$$(11) \qquad (r_2/9)^{-k}\|V\|(\chi_T^2(r_1/9,\cdot)) \ge \|V\|(\chi_T^2) - 4k\beta^2 - \alpha\beta$$

$$\ge (\nu-1/2)\underline{\underline{\beta}} - 4k\beta^2 - \alpha\beta \quad \text{and}$$

$$(12) \qquad (r_1/3)^{-k}\|V\|(\chi_T^2(r_1/3,\cdot)) \le (\nu+1/2)\underline{\underline{\beta}} + 4k\beta^2 + \alpha\beta.$$

We may choose c_{10} large enough so that either (3) is satisfied trivially, or else

(13) $$8100k\mu^2 < (\underline{\alpha}/8)9^{-k}.$$

Combining (9)-(13) with the properties of r_1 and χ yields

(14) $$9^k\|V\|\underline{B}(0,1/9) \geq (\nu-5/8)\underline{\alpha} - 4k\beta^2 - \alpha\beta - \underline{\alpha}/8,$$

(15) $$3^k\|V\|\underline{B}(0,1/3) \leq (\nu+5/8)\underline{\alpha} + 8k\beta^2 + 2\alpha\beta.$$

Therefore either

(16) $$4k\beta^2 + \alpha\beta > \underline{\alpha}/8$$

or else we may apply 5.4 with $\varepsilon = 1/8$. In the latter case, we deduce from 5.4 that there is $P < \infty$ such that

(17) $$|9^k\|V\|(\chi_T^2(1/9,\cdot)) - \nu\underline{\beta}| \leq P[\alpha^{2k/(k-2)} + \beta^2 + \mu^2].$$

Recall from 5.4 that $\alpha^{2k/(k-2)}$ is not present when $k = 1,2$. Using the cylindrical growth lemma 6.4 again,

(18) $$|9^k\|V\|(\chi_T^2(1/9,\cdot)) - \|V\|(\chi_T^2)| \leq 4k\beta^2 + \alpha\beta.$$

Combining (7), (17), and (18), we see that there is $M < \infty$ such that

(19) $$|\,\|V\|(\chi_T^2) - \nu\underline{\beta}| \leq M\sup\{\alpha^{2k/(k-2)}, \alpha\mu, \mu^2\}.$$

If $c_{10} > M$ and we let $E = \|V\|(\chi_T^2) - \nu\underline{\beta}$, then either (3) holds or

(20) $$\alpha^2 \geq \inf\{(|E|/M)^{(k-2)/k},\ |E|^2/\mu^2M^2.\}$$

If, instead, (16) holds, we infer using (7) that

(21) $$(8k+4p)\alpha\mu + 2^{1/2}\alpha^{3/2}\mu^{1/2} + 64k\rho^2\mu^2 \geq \underline{\alpha}/8.$$

If c_{10} is large enough, then either (3) holds or there is a constant $\delta > 0$ such that (21) implies

(22) $$\alpha^2 > \delta.$$

Then 6.6, (20), and (22) imply

(23) $$\delta(V,\chi_T^2)(\underline{h}(V,\cdot)) \leq -(1/4)\inf\{(|E|/M)^{(k-2)/k},\ |E^2|/\mu^2M^2,\delta\}.$$

Restoring the variable t, we have

(24) $$\overline{D}E(t) \leq -(1/4)\inf\{(|E(t)|/M)^{(k-2)/k},\ |E(t)|^2/\mu^2M^2,\delta\}.$$

Thus the maximum length of time (8) can hold and (3) not hold is

$$\Delta t = \sup\{4k\underline{\beta}^{k/2}M^{(k-2)/k},\underline{\beta}/2,\underline{\beta}/2\delta\}.$$

If $E(t_1)$ starts out greater than $c_{10}\mu^2$, then we see that $E(t) < c_{10}\mu^2$ for $t > t_1 + \Delta t$. If $E(t_1) \leq -c_{10}\mu^2$, then

$E(t_1 + \Delta t) \le -\underline{\underline{\beta}}/2$, and we can go through the above with $\nu-1$ instead of ν. Thus, we take $c_{11} = 2\Delta t$. □

Remark: This lemma can be done with $\nu \ge 2$, but it would require χ approximating the characteristic function of $\underline{\underline{B}}(0,1)$ better.

6.7. Truncated heat kernel.

We would like to exploit the close similarity between heat transfer and motion by mean curvature to show that a varifold smoothes itself out. To this end, define a truncated kernel for the heat equation as follows: Fix $T \in \underline{\underline{G}}(u,k)$. For $0 < t < 1$ and $x \in \underline{\underline{R}}^n$ let

$$(1) \qquad \psi(t,x) = \underline{\underline{c}}\, t^{-k/2}\chi_T(1/2,x)\exp[-|T(x)|^2/4t],$$

where $\underline{\underline{c}}$ is chosen so that $\int_T \psi(t,x)\,d\mathscr{H}^k x = 1.$

One may calculate that

$$(2) \qquad |\partial\psi/\partial t - \Delta\psi| < (4\rho/t)(4\pi t)^{-k/2}\exp[-1/16t].$$

6.8. Near diffusion.

Here we show that instantaneously the height of a nearly flat moving varifold is diffusing like heat. This is done by looking at the behavior of the varifold with respect to the truncated heat kernel just defined. The estimate is in terms of an upper bound in one direction, but we will later look in all directions to get a complete estimate.

Lemma: *Suppose*

(1) $V \in \mathbf{IV}_k(\mathbf{R}^n)$, $T = \mathbf{e}_1 \wedge \cdots \wedge \mathbf{e}_k \in \mathbf{G}(n,k)$,

$m \in \{k+1,\cdots,n\}$, $\tau > 0$, $0 < t < 1$,

(2) $\operatorname{spt}\|V\| \cap \mathbf{C}(T,0,9) \subset \{x \in \mathbf{R}^n : \tau < x_m < 3\tau\}$,

(3) $\theta^k(\|V\|,x) = 1$ *for* $\|V\|$ *almost all* $x \in \mathbf{C}(T,0,9)$,

(4) $\|V\|\mathbf{B}(0,1) > \boldsymbol{\alpha}/2$, *and*

(5) $\|V\|\mathbf{B}(0,3) < (3\boldsymbol{\alpha}/2)3^k$, $\|V\|\mathbf{C}(T,0,9) < 2\cdot 9^k\boldsymbol{\alpha}$.

Then there is $c_{15} < \infty$ *such that*

(6)
$$\int_{\mathbf{B}^k(0,1/3)} \max 0, \delta(V, x_m\psi(t,x-z))(\mathbf{h}(V,x)) - \|V\|(x_m \partial\psi(t,x-z)/\partial t)\}d\mathscr{H}^k z$$
$$< c_{15}t^{-1}[\tau^2 + \int_{\mathbf{B}(0,8)} |\mathbf{h}(V,x)|^2 d\|V\|x + \tau t^{-k/2}\exp(-1/16t)].$$

Proof: By the convention of 3.2, the conclusion is trivial unless the first variation of V is entirely represented by the mean curvature in $\underline{\underline{B}}(0,8)$. In the latter case, for each $z \in \underline{\underline{B}}^k(0,1/3)$, by 3.2,

$$\begin{aligned}(7)\qquad & \delta(V, x_m\psi(t,x-z))(\underline{\underline{h}}(V,x)) \\ &= -\int |\underline{\underline{h}}(V,x)|^2 x_m\psi(t,x-z)\,d\|V\|x \\ &+ \int x_m\underline{\underline{h}}(V,x)\cdot S^{\perp}(D\psi(t,x-z))\,dV(x,S) \\ &+ \int \underline{\underline{h}}(V,x)\cdot S^{\perp}(\underline{\underline{e}}_m)\psi(t,x-z)\,dV(x,S).\end{aligned}$$

The first term on the right hand side of (7) is nonnegative and thus can be neglected for conclusion (6). To estimate the second term, we use the fact that ψ depends only on $|T(x)|$ and Minkowski's inequality:

$$\begin{aligned}(8)\qquad & \left|\int x_m\underline{\underline{h}}(V,x)\cdot S^{\perp}(D\psi(t,x-z))\,dV(x,S)\right| \\ &\le 3\tau\int |\underline{\underline{h}}(V,x)|\,|D\psi(t,x-z)|\,\|S-T\|\,dV(x,S) \\ &\le 3\tau\int |\underline{\underline{h}}(V,x)|^2|D\psi(t,x-z)|\,d\|V\|x \\ &\quad + 3\tau\int \|S-T\|^2|D\psi(t,x-z)|\,dV(x,S).\end{aligned}$$

We will return to (8) later.

The third term of (7) is the significant one. By the perpendicularity of mean curvature and the definitions of 2.9,

$$(9) \qquad \int \underline{\underline{h}}(V,x)\cdot S^{\perp}(\underline{\underline{e}}_m)\psi(t,x-z)\,dV(x,S)$$

$$= \int \underline{\underline{h}}(V,x)\cdot \underline{\underline{e}}_m\psi(t,x-z)\,dV(x,S)$$

$$= -\delta V(\underline{\underline{e}}_m\psi(t,x-z))$$

$$= -\int \underline{\underline{e}}_m\cdot S(D\psi(t,x-z))\,dV(x,S).$$

We next estimate this integral using the Lipschitz approximations $f\colon T \to T^{\perp}$ and $F\colon T \to \underline{\underline{R}}^n$ constructed in 5.4 for $\nu = 1$ and $p = 2$. Let X and Y be as in 5.4, and note that since $\nu = 1$ we may take $\mathrm{Lip}(f) = 1$. Estimating as in 5.6 (39)-(41),

$$(10) \qquad \int \underline{\underline{e}}_m\cdot S(D\psi(t,x-z))\,dV(x,S) - \int_T Df_m(y)\cdot D\psi(t,y-z)\,d\mathscr{H}^k y$$

$$= \int_{\underline{\underline{B}}(0,1)\sim X} \underline{\underline{e}}_m\cdot S(D\psi(t,x-z))\,dV(x,S)$$

$$+ \int_Y e_m\cdot \mathrm{image}\ DF(y)(D\psi(t,y-z))\,|\Lambda_k DF(y)| - Df_m(y)\cdot D\psi(t,y-z)\,d\mathscr{H}^k y$$

$$+ \int_{\underline{\underline{B}}^k(0,1)\sim Y} Df_m(y)\cdot D\psi(t,y-z)\,d\mathscr{H}^k y$$

$$\leq \int_{\underline{\underline{B}}(0,1)\sim X} |D\psi(t,y-z)|\,d\|V\|x$$

$$+ \underline{\underline{c}}_2 \int_X \|S-T\|^2\,|D\psi(t,x-z)|\,dV(x,S)$$

$$+ \int_{\underline{\underline{B}}^k(0,1)\sim Y} |D\psi(t,y-z)|\,d\mathscr{H}^k y,$$

where $\underline{\underline{c}}_2$ is from [AW1 8.14]. These error estimates are not simplified further because later they will be integrated with respect to z.

Since f is Lipschitz and ψ has compact support, we can integrate by parts:

$$(11) \qquad \int_T Df_m(y)\cdot D\psi(t,y-z)\,d\mathscr{H}^k y$$

$$= -\int_T f_m(y)\,\Delta\psi(t,y-z)\,d\mathscr{H}^k y.$$

We then make a similar set of estimates to get back to the varifold:

$$(12) \qquad - \int_T f_m(y)\,\Delta\psi(t,y-z)\,d\mathscr{H}^k y$$

$$+ \int x_m \Delta\psi(t,x-z)\,d\|V\|x$$

$$= \int_{\underline{\underline{B}}(0,1)\sim X} x_m \Delta\psi(t,y-x)\,d\|V\|x$$

$$+ \int_Y f_m(y)\,\Delta\psi(t,y-z)\,[\,|\Lambda_k DF(y)|-1]\,d\mathscr{H}^k y$$

$$+ \int_{\underline{\underline{B}}^k(0,1)\sim Y} f_m(y)\,\Delta\psi(t,y-z)\,d\mathscr{H}^k y$$

$$\leq 3\tau \int_{\underline{\underline{B}}(0,1)\sim X} |\Delta\psi(t,x-z)|\,d\|V\|x$$

$$+ 3\tau\,\underline{\underline{c}}_2 \int_X \|S-T\|^2\,|\Delta\psi(t,x-z)|\,dV(x,S)$$

$$+ 3\tau \int_{\underline{\underline{B}}^k(0,1)\sim Y} |\Delta\psi(t,y-z)|\,d\mathscr{H}^k y.$$

$$(13)\quad \int x_m \Delta\psi(t,x-z)\,d\|V\|x - \int x_m \partial\psi(t,x-z)/\partial t\, d\|V\|x$$

$$\le 3\tau\,|\Delta\psi-\partial\psi/\partial t|\,\|V\|\underline{\underline{C}}(T,0,1)$$

$$< 3\tau(4\rho/t)\underline{\underline{\sigma}}t^{-k/2}\exp[-1/16t]\underline{\underline{\alpha}}3^{k+1}.$$

To facilitate integrating these estimates over z, define

$$(14)\quad \zeta_1(t,x) = \int_{\underline{\underline{B}}^k(0,1/3)} |D\psi(t,x-z)|\,d\mathscr{H}^k z \quad \text{and}$$

$$(15)\quad \zeta_2(t,x) = \int_{\underline{\underline{B}}^k(0,1/3)} |\Delta\psi(t,z-x)|\,d\mathscr{H}^k z.$$

Therefore, adding (8), (10)-(13), and integrating over z,

$$(16)\quad \int_{\underline{\underline{B}}^k(0,1/3)} \max\{0,\delta(V,x_m\psi(t,x-z))(\underline{\underline{h}}(V,x)) - \|V\|(x_m\partial\psi(t,x-z)/\partial t)\}\,d\mathscr{H}^k z$$

$$\le 3\tau \int |\underline{\underline{h}}(V,x)|^2\zeta_1(t,x)\,d\|V\|x$$

$$+ \int_{\underline{\underline{B}}(0,1)\sim X} \zeta_1(t,x) + 3\tau\zeta_2(t,x)\,d\|V\|x$$

$$+ \underline{\underline{c}}_2 \int_X \|S-T\|^2(\zeta_1(t,x) + 3\tau\zeta_2(t,x))\,dV(x,S)$$

$$+ \int_{\underline{\underline{B}}^k(0,1)\sim Y} \zeta_1(t,y) + 3\tau\zeta_2(t,y)\,d\mathscr{H}^k y$$

$$+ 3\tau(4\rho/t)\underline{\underline{\sigma}}t^{-k/2}\exp[-1/16t]\underline{\underline{\alpha}}3^{k+1}\mathscr{H}^k\underline{\underline{B}}^k(0,1/3).$$

One may compute that

(17) $$\sup\{\zeta_1(t,x) : x \in R^n\} \le kt^{-1/2} \quad \text{and}$$

(18) $$\sup\{\zeta_2(t,x) : x \in \underline{\underline{R}}^n\} < kt^{-1}.$$

Therefore (16) becomes

(19) $$\int_{\underline{\underline{B}}^k(0,1/3)} \max\{0, \delta(V, x_m\psi(t,x-z))(\underline{\underline{h}}(V,x)) - \|V\|(x_m \partial\psi(t,x-z)/\partial t)\} d\mathscr{H}^k z$$

$$\le 3\tau k t^{-1/2} \int_{\underline{\underline{B}}(0,1)} |\underline{\underline{h}}(V,x)|^2 d\|V\|x$$

$$+ (kt^{-1/2} + 3\tau k t^{-1})[\|V\|(\underline{\underline{B}}(0,1) \sim X) + \mathscr{H}^k(\underline{\underline{B}}^k(0,1) \sim Y)]$$

$$+ \underline{\underline{c}}_2(kt^{-1/2} + 3\tau k t^{-1}) \int_{\underline{\underline{B}}(0,1)\times\underline{\underline{G}}(n,k)} \|S-T\|^2 dV(x,S)$$

$$+ 3\tau(4\rho/t)\underline{\underline{c}}t^{-k/2}\exp[-1/16t]3\underline{\underline{\alpha}}^2.$$

From the Lipschitz approximation theorem 5.4

(20) $$\|V\|(\underline{\underline{B}}(0,1) \sim X) + \mathscr{H}^k(\underline{\underline{B}}^k(0,1) \sim Y) < p\Big[\int_{\underline{\underline{B}}(0,7)} |\underline{\underline{h}}(V,x)|^2 d\|V\|x$$

$$+ \int_{\underline{\underline{B}}(0,7)} \|S-T\|^2 dV(x,S)$$

$$+ \int_{\underline{\underline{B}}(0,7)} |T^{\perp}(x)|^2 d\|V\|x\Big].$$

Also, from 5.5 (6) with $\phi = \chi_T(8,\cdot)$ and Minkowski's inequality,

$$(21) \quad \int_{\underline{\underline{B}}(0,7)} \|S-T\|^2 dV(x,S) \le 16(\rho/8)^2 \int_{\underline{\underline{B}}(0,8)} |T^{\perp}(x)|^2 d\|V\|x$$

$$+ 2[\int_{\underline{\underline{B}}(0,8)} |\underline{\underline{h}}(V,x)|^2 d\|V\|x \int_{\underline{\underline{B}}(0,8)} |\dot{T}^{\perp}(x)|^2 d\|V\|x]^{1/2}$$

$$\le \rho^2 \tau^2 \|V\| \underline{\underline{B}}(0,8) + \int_{\underline{\underline{B}}(0,8)} |\underline{\underline{h}}(V,x)|^2 d\|V\|x.$$

Since $t < 1$, we have $t^{-1/2} < t^{-1}$, so we see from (5), (19), (20), and (21) that there is a constant $c_{15} < \infty$ such that the conclusion of the lemma holds. □

6.9. Flattening out.

We apply the previous lemma to a moving varifold to show that if the varifold is reasonably flat on a certain scale to start with, then later it is much flatter on a much smaller scale. This result is somewhat like that of 5.6, but with time thrown in. However the proof is much different, relying on the heat analogy rather than blowing up. In 5.6, the curvature had to be small compared to the roughness, but here it is large, although we are able to put a bound on the ratio. This is the lemma where the unit density hypothesis is critical.

We say that a varifold V_t moving by its mean curvature has unit density if $\Theta^k(\|V_t\|,x) = 1$ for $\|V_t\|$ almost all $x \in \underline{\underline{R}}^n$ for almost all $t > 0$.

Lemma: *For any* $\varepsilon > 0$ *there exists* $\theta(\varepsilon) > 0$ *such that if* $0 < R \leq \Theta(\varepsilon)$ *then there exist* $0 < \eta_1 < 1$ *with the following property*:

If V_t *is a unit density integral varifold moving by its mean curvature*,

(1) $$T = \underline{\underline{e}}_1 \wedge \cdots \wedge \underline{\underline{e}}_k \in \underline{\underline{G}}(n,k), \quad 0 < \tau < \eta_1,$$

$$0 \leq t_0 < t_1 < \infty;$$

(2) $$\text{spt}\|V_t\| \cap \underline{\underline{C}}(T,0,9) \subset \{x: |T^{\perp}(x)| < \tau\},$$

(3) $$\underline{\beta}/2 \leq \|V\|(\chi_T^2) \leq 3\underline{\beta}/2,$$

(4) $$\|V_t\|\underline{\underline{B}}(0,3) \leq (3\underline{\alpha}/2)3^k, \|V_t\|\underline{\underline{C}}(T,0,9) < 2\cdot 9^k\underline{\alpha},$$

for almost all $t \in [t_0,t_1]$;

(5) c_{11} is as in 6.7 and $t_0 + c_{11} + 1 < s_0 < t_1 - c_{11} - 1$;

then there exists $A \in \underline{\underline{A}}(n,k)$ *such that if* $A = T^* + a$, $T^* \in \underline{\underline{G}}(n,k)$, *and* $a \in T^{\perp}$, *then*

(6) $$\|T^*-T\| \leq 2\tau, \quad |a| < 2\tau,$$

(7) $$\text{spt}\|V_t\| \cap \underline{\underline{C}}(T^*,a,R) \subset \{x: \text{dist}(x,A) < R^{2-\varepsilon}\tau\},$$

(8) $$(\underline{\beta}/2) \leq (R/9)^{-k}\|V_t\|(\chi_{T^*}^2(R/9,\cdot)) \leq 3\underline{\beta}/2, \quad \textit{and}$$

(9) $\|V_t\|\underline{\underline{B}}(a,R/3) \leq (3\underline{\underline{\alpha}}/2)(R/3)^k, \|V_t\|\underline{\underline{C}}(T^*,a,R) \leq 2\cdot\underline{\underline{\alpha}}R^k$

for $s_0 \leq t \leq s_0 + 4(c_{11}+1)R^2$.

Proof: We may assume that (2)-(4) hold for all $t \in [t_0,t_1]$ because we will be concerned with integrals over t. Applying 6.6 with $\mu^2 = 2\underline{\underline{\alpha}}\tau^2$, we see from (3) that

$$(10) \qquad -c_{10}^2\mu^2 < \|V_t\|(\chi_T^2) - \underline{\underline{\beta}} < c_{10}^2\mu^2$$

for $t_0 + c_{11} \leq t \leq t_1 - c_{11}$. This means that there is a limited amount of mass to be lost in this time interval, and hence a limited amount of mean curvature. In fact, defining

$$(11) \qquad \alpha(t)^2 = \int |\underline{\underline{h}}(V,x)|^2\chi_T^2(x)\,d\|V_t\|x,$$

by 6.5 (3) we must have

$$(12) \qquad \int_{s-1}^{s} -\alpha(t)^2/2 + 32\rho^2\mu^2 dt \geq \int_{s-1}^{s} \bar{D}\|V_t\|(\chi_T^2)\,dt$$
$$> \|V_s\|(\chi_T^2) - \|V_{s-1}\|(\chi_T^2)$$
$$> -2c_{10}\mu^2.$$

Hence

$$(13) \qquad \int_{s-1}^{s} \alpha(t)^2 dt < 4(c_{10} + 16\rho^2)\mu^2,$$

where the relationship of s to s_0 will be defined at eq. (24).

To apply the previous lemma, 6.8, let $p = \tau^{1/2}$ and

$q \approx 36R^{\varepsilon}-p$, where q will be pinned down later. Let y be a unit vector in $T^{\perp}$, and assume $y = \underline{\underline{e}}_m$. Define $W_t = \underline{\tau}(2\tau\underline{\underline{e}}_m)_{\#}V_t$, so that

$$\text{(14)} \qquad \operatorname{spt}\|W\| \cap \underline{\underline{C}}(T,0,1) \subset \{x: \tau < x_m < 3\tau\}.$$

Adjusting for the different scale in 6.8, we infer from 3.6 and 6.10 that

$$\text{(15)} \qquad \int_{\underline{\underline{B}}^k(0,1/27)} \max\{0, \|W_s\|(\psi(p,x-z)) - \|W_{s-q}\|(\psi(p+q,x-z))\}d\mathscr{H}^k z$$

$$\leq \int_{\underline{\underline{B}}^k(0,1/27)} \int_{s-q}^{s} \max\{0, \overline{D}\|W_t\|(\psi(p+s-t,x-s))\}dt\,d\mathscr{H}^k z$$

$$\leq \int_{s-q}^{s} \int_{\underline{\underline{B}}^k(0,1/27)} \max\{0, \delta(W_t,\psi(p+s-t,x-z))(\underline{h}(W,x))$$

$$+ \|W_t\|(\partial\psi(p+s-t,x-z)/\partial t)\}d\mathscr{H}^k z\,dt$$

$$\leq \int_{s-q}^{s} c_{15}(p+s-t)^{-1}[\tau^2+\alpha(t)^2+\tau(p+s-t)^{-k/2}\exp(-1/16(p+s-t))]dt,$$

where the scaling factor involved in applying 6.8 has been absorbed into c_{15}. Using (13) and $p \leq p+s-t \leq p+q$, we have

$$\text{(16)} \qquad \int_{\underline{\underline{B}}^k(0,1/27)} \max\{0, \|W_s\|(\psi(p,x-z))$$

$$- \|W_{s-q}\|(\psi(p+q,x-z))\}d\mathscr{H}^k z$$

$$\leq c_{15}p^{-1}[1 + 4(c_{10}+16\rho^2)]\tau^2$$

$$+ c_{15}32(p+q)^{(2-k)/2}\exp(-1/16(p+q))\tau.$$

Define an affine map $L: T \to \mathbb{R}$ by

$$(17) \qquad L(z) = \|W_{s-q}\|(x_m \psi(p+q,x) - x_m z \cdot D\psi(p+q,x)).$$

To calculate how much W deviates above the plane $x_m = L(T(x))$, we first compute how far $\|W_{s-q}\|(x_m \psi(p+q,x-z))$ deviates. By Taylor's formula with remainder,

$$(18) \qquad \psi(p+q,x-z) = \psi(p+q,x) - z \cdot D\psi(p+q,x)$$
$$+ \int_0^1 (1-\theta)(z \otimes z) \cdot D^2\psi(p+q,x-\theta z)\, d\theta.$$

One may calculate from 6.7 that for $W \in T$

$$(19) \qquad D\psi(t,w) = [D^2\chi(1/2,w) - D\chi(1/2,w) \otimes w/t - \chi(1/2,w)T/2t$$
$$+ \chi(1/2,w)(w \otimes w)/4t^2](4\pi t)^{-k/2} \exp(-|w|^2/4t).$$

Therefore,

$$(20) \qquad \|W_{s-q}\|(x_m \psi(p+q,x-z)) - L(z)$$
$$\leq 3\tau \int_0^1 (1-\zeta) \|W_{s-q}\|((z \otimes z) \cdot D^2\psi(p+q,x-\zeta z))\, d\zeta$$
$$\leq 3\tau |z|^2 4\rho (4\pi(p+q))^{-k/2} \exp[-1/16(p+q)] \|W_{s-q}\| \underline{\underline{C}}(T,0,1)$$
$$+ 3\tau |z|^2 (2\rho/(p+q))(4\pi(p+q))^{-k/2} \exp[-1/16(p+q)] \|W_{s-q}\| \underline{\underline{C}}(T,0,1)$$
$$+ 3\tau \int_0^1 (1-\zeta)(|z|^2/2(p+q))$$
$$\cdot \int [1 + (|T(x)|^2 + |z|^2)/(p+q)] \psi(p+q,x-\zeta z)\, d\|W_{s-q}\| x\, d\zeta.$$

We may find q with $36R^{\varepsilon} < p+q < 72R^{\varepsilon}$ and

(21) $$\alpha^2(s-q) < 8(c_{10}+16\rho^2)\mu^2/q.$$

Then we may use the Lipschitz approximation theorem 5.4 to see that there is a constant c_{18} such that when $|T(z)|^2 < p+q$ we have

(22) $$\int [1+(|T(x)|^2+|z|^2)/(p+q)]\psi(p+q,x-\theta z)\,d\|W_{s-q}\|x$$
$$\leq 3 + c_{18}(p+q)^{-(k+2)/2}\tau^2.$$

Hence, in case (19) holds,

(23) $$\|W_{s-q}\|(x_m\psi(p+q,x-z))-L(z)$$
$$\leq 3\tau(|z|^2/2(p+q))(3+c_{18}(p+q)^{-(k+2)/2}\tau^2)$$
$$< 6\tau|z|^2/(p+q) \quad \text{if} \quad c_{18}(p+q)^{-(k+2)/2}\tau^2 < 1.$$

We shall take τ small enough compared to R so that (23) does hold.

We have already seen that $\|W_{s-q}\|(x_m\psi(p+q,x-z))$ approximates $\|W_s\|(x_m\psi(p,x-z))$, so let's see how the graph of $\|W_s\|(x_m\psi(p,x-z))$ as a function of z approximates W_s. This is where the unit density hypothesis is critical.

By (13), we may find s with $s_0-2(c_{11}+1)R^2 < s < s_0$ and

(24) $$\alpha(s)^2 < 4(c_{10}+16\rho^2)\mu^2R^{-2}.$$

Then letting $f\colon T \to T^{\perp}$ be the Lipschitz approximation constructed in 5.4, we see using (24), 5.5, and (14) that there is a constant c_{19} such that

(25) $$\|W_s\|\{x \in \underline{\underline{C}}(T,0,1/27) : x_m > L(T(x)) + 6\tau|T(x)|^2/(p+q)\}$$
$$< \mathscr{H}^k\{z \in \underline{\underline{B}}^k(0,1/27) : f_m(z) > L(z) + 6\tau|z|^2/(p+q)\}$$
$$+ c_{19}\tau^2 R^{-2} \quad \text{and}$$

(26) $$\int_{\underline{\underline{B}}(0,1/27)} |\|W_s\|(x_m\psi(p,x-z)) - \int_T f_m(y)\psi(p,y-z)\,d\mathscr{H}^k y|\,d\mathscr{H}^k z$$
$$< c_{19}\tau^2 R^{-2}.$$

Estimate (26) is similar to those made in 6.8. Next, we may calculate

(27) $$\int_{\underline{\underline{B}}(0,1/27)} |f_m(z) - \int_T f_m(y)\psi(p,y-z)\,d\mathscr{H}^k y|\,d\mathscr{H}^k z$$
$$\leq \int_{\underline{\underline{B}}^k(0,1/27)} |\int_0^p \int_T \partial\psi(t,y-z)/\partial t f_m(y)\,d\mathscr{H}^k y dt|\,d\mathscr{H}^k z$$
$$\leq \int_{\underline{\underline{B}}^k(0,1/27)} |\int_0^p \int_T \Delta\psi(t,y-z) f_m(y)\,d\mathscr{H}^k y dt|\,d\mathscr{H}^k z$$
$$+ 3\tau\underline{\underline{\alpha}}^2 \int_0^p |\partial\psi/\partial t - \Delta\psi|\,dt$$
$$\leq \int_{\underline{\underline{B}}^k(0,1/27)} |\int_0^p \int_T D\psi(t,y-z)\cdot Df_m(y)\,d\mathscr{H}^k y dt|\,d\mathscr{H}^k z$$
$$+ 3\tau\underline{\underline{\alpha}}^2 \int_0^p (4\rho/t)(4\pi t)^{-k/2}\exp[-1/16t]\,dt$$

$$\leq \int_0^p \int_T \zeta_1(t,y)\,|Df_m(y)|\,d\mathscr{H}^k y\,dt + p\tau \quad \text{where} \quad \zeta_1 \quad \text{is as in 6.8 (14),}$$

$$\leq \int_0^p \int_T kt^{-1/2}|Df_m(y)|\,d\mathscr{H}^k dt + p\tau$$

$$\leq 2kp^{1/2} \int_{\underline{\underline{B}}^k(0,1/27)} |Df_m(y)|\,d\mathscr{H}^k + p\tau$$

$$\leq c_{19}p^{1/2}\tau R^{-1} + p\tau.$$

Combining (25), (27), (26), (16), and (23), we get

$$(28) \quad \int_{\underline{\underline{B}}(0,1/27)} \max\{0, x_m - L(T(x)) - 6\tau|T(x)|^2/(p+q)\}\,d\|W_s\|x$$

$$\leq \int_{\underline{\underline{B}}^k(0,1/27)} \max\{0, f_m(z) - L(z) - 6\tau|z|^2/(p+q)\}\,d\mathscr{H}^k z + 3\tau\cdot c_{19}\tau^2R^{-2}$$

$$\leq \int_{\underline{\underline{B}}^k(0,1/27)} |f_m(z) - \int_T f_m(y)\psi(p,y-z)\,d\mathscr{H}^k y|\,d\mathscr{H}^k z$$

$$+ \int_{\underline{\underline{B}}^k(0,1/27)} |\int_T f_m(y)\psi(p,y-z)\,d\mathscr{H}^k y - \|W_s\|(x_m\psi(p,x-z))|\,d\mathscr{H}^k z$$

$$+ \int_{\underline{\underline{B}}^k(0,1/27)} \max\{0, \|W_s\|(x_m\psi(p,x-z)) - \|W_{s-q}\|(x_m\psi(p+q,x-z))\}\,d\mathscr{H}^k z$$

$$+ \int_{\underline{\underline{B}}^k(0,1/27)} \max\{0, \|W_{s-q}\|(x_m\psi(p+q,x-z)) - L(z) - 6\tau|z|^2/(p+q)\}\,d\mathscr{H}^k z$$

$$+ 3c_{19}\tau^3R^{-2}$$

$$\leq c_{19}p^{1/2}\tau R^{-1} + p\tau + c_{19}\tau^2R^{-2}$$

$$+ c_{15}[p^{-1}[1+4(c_{10}+16\rho^2)]\tau^2$$

$$+ 32(p+q)^{(2-k)/2}\exp(-1/16(p+q))\tau]$$

$$+ 0 + 3c_{19}\tau^3 R^{-2}.$$

For convenience, write this last quantity as $\delta\tau$.

Now consider the ball $\underline{\underline{B}}(b,R_0)$ that has radius $R_0 = (p+q)/12\tau$ and center b in the plane spanned by T and $\underline{\underline{e}}_m$ and is tangent to the graph of the paraboloid

(29) $$G = \{x \in \underline{\underline{R}}^n\colon x_m = L(T(x))+6\tau|T(x)|^2/(p+q)\}$$

at the point $(0,L(0)) \in \underline{\underline{R}}^k \times \underline{\underline{R}}^{n-k}$. Note that $\underline{\underline{B}}(b,R_0)$ is is entirely "above" G.

We want to apply the clearing out lemma 6.3 to $\underline{\underline{B}}(b,R_0)$. Letting

(30) $$\phi(x) = \begin{cases} 1 - |x-b|^2/R_0^2 & \text{for} \quad 0 \le |x-a| \le R_0 \\ 0 & \text{for} \quad |x-a| \ge R_0, \end{cases}$$

note that

(31) $$\phi^3(x) = R_0^{-6}(R_0+|x-b|)^3(R_0-|x-b|)^3$$
$$\le 8R_0^{-3}4\tau^2(R_0-|x-b|) \quad \text{and}$$

(32) $$R_0-|x-b| < x_m-L(T(x))+6\tau^2|T(x)|^2/(p+q)$$

for $x \in \underline{\underline{B}}(b,R_0) \cap \operatorname{spt}\|V_t\| \cap \underline{\underline{C}}(T,0,1)$. Even though it may be that $\underline{\underline{B}}(b,R_0)$ is not entirely contained in $\underline{\underline{C}}(T,0,1)$, we can check that $\underline{\underline{B}}(b,R_0) \cap \{x\colon x_m < 3\tau\}$ is in $\underline{\underline{C}}(T,0,1)$. We may thus modify ϕ to vanish outside $\underline{\underline{C}}(T,0,1)$ without affecting the following calculations. From (28), (30), and (31) we find that

$$(33) \qquad \|W_s\|(\phi^3) \leq 32R_0^{-3}\delta\tau^3$$
$$\leq 32\cdot(12/(p+q))^{k+3}\delta\tau^{k+6}R_0^k.$$

Hence, by the clearing out lemma 6.3 with $m' = 3$, we have

$$(34) \qquad \|W_{s+\Delta s}\|\underline{\underline{B}}(b,R_1) = 0$$

when $\Delta s \geq \Delta s_0$, where

$$(35) \qquad \Delta s_0 = c(3)[32\cdot(12/(p+q))^{k+3}\tau^{6+k}\delta]^{2/(k+6)}[(p+q)/12\tau]^2$$
$$= c(3)\,32^{2/(k+6)}(12/(p+q))^{-6/(k+6)}\delta^{2/(k+6)}, \quad \text{and}$$

$$R_0-R_1 = 4k\Delta s/(R_0+R_1) \geq 2k\Delta s/R_0$$
$$\geq c(3)\,32^{2/(k+6)}(12/(p+q))^{k/(k+6)}\delta^{2/(k+6)}\tau$$

We want to have τ, p, and q so that

$$(36) \qquad \Delta s_0 < R^2$$

and

$$(37) \qquad R_0-R_1 < R^{2-\varepsilon}\tau/3$$

when

(38) $$\Delta s < 8(c_{11}+1)R^2.$$

We also want for (23) that

(39) $$\tau^2 < c_{18}^{-1}(p+q)^{(k+2)/2}.$$

With $p = \tau^{1/2}$ and $p+q \approx 36R^{\varepsilon}$, we see that the only term of δ in (26) that does not contain a positive power of τ is the exponential term. For R^{ε} small enough, this term is small compared to the relevant powers of R. Then τ can be chosen small enough compared to R so that δ is small enough that (36), (37), and (38) hold.

Now go back to V_t. Let the sought-for $A \in \underline{\underline{A}}(n,k)$ be the graph of the affine maps $L^*: T \to T^{\perp}$ defined by

$$L^*(z) = \|V_{s-q}\|(T^{\perp}(x)\psi(p+q,x) - T^{\perp}(x)z\cdot D\psi(p+q,x)).$$

We see that (6) is easily satisfied. To check (7), note that

$$|L^*(z)+2\tau e_m - L(z)| \leq 2\tau|\|V_{s-q}\|(\psi(p+q,x))-1|$$

$$+ 2\tau|z||\|V_{s-q}\|(D\psi(p+q,x))|.$$

We use the Lipschitz approximation lemma 5.4 as above to see that

$$\|V_{s-q}\|(\psi(p+q,x)) - \int_T \psi(p+q,y)\,d\mathscr{H}^k y < c_{18}(p+q)^{-k/2}\tau^2$$

and

$$|\|V_{s-q}\|(D\psi(p+q,x))| - \int_T D\psi(p+q,y)\,d\mathscr{H}^k y < c_{18}(p+q)^{(-k-1)/2}\tau^2,$$

which shows that

$$|L^*(z)+2\tau e_m-L(z)| < 2c_{18}(p+q)^{(-k-1)/2}\tau^2$$

$$< R^{2-\varepsilon}\tau/3$$

for small τ.

Since $\mathbf{e}_m$ was in an arbitrary direction, and the choices made of times and such did not depend on $\mathbf{e}_m$, we put every thing together to conclude that for $x \in \operatorname{spt}\|V_t\|$ and for $s+\Delta s_0 < t < s+\Delta s_0+8(c_{11}+1)R^2$

$$|T(x)-L^*(T(x))| \leq 2[6\tau|T(x)|^2/(p+q)]$$

$$+ R_0-R_1 + R^{2-\varepsilon}\tau/3.$$

Hence for $|T(x)| < R$,

$$|T(x)-L^*(T(x))| \leq (R^{2-\varepsilon}\tau + R^{2-\varepsilon}\tau + R^{2-\varepsilon}\tau)/3.$$

Finally, we verify (7) and (8). The various upper bounds clearly hold for V_s, and by the film popping lemma 6.6 they remain true. If any of the lower bounds were violated at time t, then 6.6 would lead to violation of the lower bounds in the hypotheses at time $t + 2c_{11}$. □

6.10. Infinite differentiability

Here we show that a nearly flat varifold becomes smooth after a little time. This is done by using the previous theorem inductively to get $\operatorname{spt}\|V_t\|$ to be the graph of a Hölder continuously differentiable function, and then using the near-diffusion lemma 6.8 again to get the function to have continuous second derivatives and to be a solution of the non-parametric quasi-linear parabolic partial differential equation mentioned in 3.1. Standard P.D.E. theory then gives the infinite differentiability.

<u>Theorem</u>: <u>There are constants</u> $c_{21} < \infty$ <u>and</u> $\tau_0 > 0$, <u>such that</u>:

<u>If</u> V_t <u>is a unit density integral varifold moving by its mean curvature</u>,

(1) $T \in \underline{\underline{G}}(n,k)$, $a \in \underline{\underline{R}}^n$, $0 \leq t_0 < t_1 < \infty$, $0 < R < \infty$;

(2) $\operatorname{Spt}\|V_t\| \cap \underline{\underline{B}}(a,R) \subset \{x: \operatorname{dist}(x,T+a) < \tau_0 R\}$,

(3) $$\underline{\underline{\beta}}/2 \leq (R/9)^{-k}\|V_t\|(\chi_T^2(R/9,\cdot)) \leq 3\underline{\underline{\beta}}/2,$$

(4) $$\|V_t\|\underline{\underline{B}}(a,R/3) \leq (3\underline{\underline{\alpha}}/2)(R/3)^k \text{ and}$$

$$\|V_t\|\underline{\underline{C}}(T,a,R) < 2\underline{\underline{\alpha}}R^k$$

<u>for all</u> $t \in [t_0,t_1]$, <u>then</u>

$$\{(t,x) \in (t_0+c_{21}R^2, t_1-c_{21}R^2) \times \underline{\underline{U}}(a,R/3): x \in \operatorname{spt}\|V_t\|\}$$

<u>is an infinitely differentiable manifold</u>.

Proof: We may assume that $T = \underline{\underline{e}}_1 \wedge \cdots \wedge \underline{\underline{e}}_k$, $a = 0$, $t_0 = 0$, and $R = 3$.

Let $\varepsilon > 0$ be arbitrary, and let $\theta = \theta(\varepsilon)$ be as in 6.9. Let $y \in \underline{\underline{B}}^k(0,1)$. We can apply 6.9 to V_t in $\underline{\underline{C}}(T,y,1) \cap \underline{\underline{B}}(0,3)$ because of the boundedness of $\operatorname{spt}\|V_t\|$ given by (2). Applying 6.9 repeatedly, with slightly tilted and every smaller cylinders of radius θ^m around y, we find that if

$$(6) \qquad (c_{11}+1)\sum_{m=0}^{\infty}(\theta^m)^2 < t < t_2-(c_{11}+1)\sum_{m=0}^{\infty}(\theta^m)^2$$

then there are $A_m(t,y) \in \underline{\underline{A}}(n,k)$ and $F(t,y) \in T^{-1}(y)$ such that

$$(7) \quad \operatorname{spt}\|V_s\| \cap \underline{\underline{C}}(T,y,\theta^m) \cap \underline{\underline{B}}(0,3) \subset \{x\colon \operatorname{dist}(x,A_m(t,y)) < \theta^{(2-\varepsilon)m}\tau_0\}$$

for $t-\theta^{2m} < s < t+\theta^{2m}$, and

$$(8) \qquad A_\infty(t,y) = \lim_{m\to\infty} A_m(t,y)$$

$$= \operatorname{Tan}^k(\operatorname{spt}\|V_t\|, F(t,y)) + F(t,y).$$

We take $c_{21} = (c_{11}+1)\sum_{m=0}^{\infty}(\theta^m)^2$.

Let $f(t,y) = T^{\perp}(F(t,y))$. Clearly, (7) shows that f is differentiable in y and Hölder continuous in t with exponent $1-\varepsilon/2$.

Now we establish a little more differentiability for f. If $A_m(t,y) = T_m(t,y) + a_m(t,y)$ with $T_m(t,y) \in \underline{\underline{G}}(n,k)$ and $a_m(t,y) \in T^{-1}(y)$, then by 6.9 (6),

(9) $\|T_m(t,y)-T_{m+1}(t,y)\| < 2\theta^{m(1-\varepsilon)}\tau_0,$

(10) $|a_m(t,y)-a_{m+1}(t,y)| < 2\theta^{(2-\varepsilon)m}\tau_0.$

Consider $y_1, y_2 \in \underline{\underline{B}}^k(0,1)$. Suppose $m \in \underline{\underline{N}}$ is such that

(11) $\theta^{m+1}/2 \le |y_1-y_2| < \theta^m/2.$

Then since (7) holds for y_1 and y_2,

(12) $\text{dist}(x,A_m(t,y)),\ \text{dist}(x,A_m(t,y_2)) < \theta^{(2-\varepsilon)m}\tau_0$

for $x \in \text{spt}\|V_t\| \cap \underline{\underline{C}}(T_1(y_1+y_2)/2, \theta^m/2) \cap B(0,3)$.

Hence

(13)
$$\|T_m(t,y_1)-T_m(t,y_2)\| < \theta^{(2-\varepsilon)m}\tau_0/(\theta^m/2)$$
$$< 2\theta^{(1-\varepsilon)m}\tau_0$$
$$< 4\tau_0\theta^{\varepsilon-1}|y_1-y_2|^{1-\varepsilon}.$$

Then by (9),

(14)
$$\|T_\infty(t,y_1)-T_\infty(t,y_2)\| < \|T_\infty(t,y_1)-T_m(t,y_1)\|$$
$$+ \|T_m(t,y_1)-T_m(t,y_2)\|+\|T_m(t,y_2)-T_\infty(t,y_2)\|$$
$$< 4\tau_0\theta^{-1/2}|y_1-y_2|^{1/2} + 2\tau_0\sum_{q=m}^{\infty}\theta^{(1-\varepsilon)q-1}$$
$$< 4\tau_0\theta^{-1/2}|y_1-y_2|^{1/2} + 2\tau_0\theta^{(1-\varepsilon)m-1}(1-\theta^{1-\varepsilon})^{-1}$$
$$< 4\tau_0\theta^{-1/2}|y_1-y_2|^{1/2} + 4\tau_0\theta^{\varepsilon-2}|y_1-y_2|^{1-\varepsilon}.$$

Hence $f(t,y)$ is Hölder continuously differentiable in y with Hölder exponent $1-\varepsilon$.

Likewise, we can show that $Df(t,y)$ is Hölder continuous in t. If $y \in \underline{\underline{B}}^k(0,1)$, $9c_{21} < t_2 < t_3 < t_1 - 9c_{21}$ and

(15) $$\theta^{2(m+1)} < |t_2 - t_3| < \theta^{2m},$$

then we may assume that $A_m(t_2,y) = A_m(t_3,y)$, and therefore by (9) again

(16) $$\|T_\infty(t_2,y) - T_\infty(t_3,y)\| \leq 4\tau_0 \theta^{(1-\varepsilon)m-1}$$
$$\leq 4\tau_0 \theta^{\varepsilon-2} |t_2 - t_3|^{(1-\varepsilon)/2}.$$

Now we re-examine the error estimates for near-diffusion as in 6.8 using our much improved smoothness. Letting τ, p, q, m and z, serve the same role as in 6.8, and noting that the Lipschitz approximations are exact, we can extract from 6.8 (8), (10), (12), (13) the estimate

(18) $$\delta(V, x_m \psi(t,x-z))(\underline{\underline{h}}(V,x)) - \|V\|(x_m \partial\psi(t,x-z)/\partial t)$$
$$\leq 3\tau \int \|S-T\|^2 |x-z|^2/t^2 \psi(t,x-z)\, dV(x,S)$$
$$+ \underline{\underline{c}}_2 \int \|S-T\|^2 |D\psi(t,x-z)|\, dV(x,S)$$
$$+ 3\tau \underline{\underline{c}}_2 \int \|S-T\|^2 |\Delta\psi(t,x-z)|\, dV(x,S)$$
$$+ 3\tau(4\rho/t)\, \underline{\underline{c}}\, t^{-k/2} \exp[-1/16t]\, \underline{\underline{\alpha}} 3^{k+1}.$$

Take a specific time t_2 and a radius $R > 0$. Re-orient everything so that $f(t_2,0) = 0$ and $Df(t_2,0) = 0$. Let M be a general purpose constant. Then, from the first part of this theorem, we can take

$$(19)\qquad x_m \leq M|t-t_2|^{1-\varepsilon/2} + M|T(x)|^{2-\varepsilon} \text{ and}$$

$$\|S-T\| < M|T(x)|^{1-\varepsilon} + M|t-t_2|^{(1-\varepsilon)/2}$$

for $\|V_t\|$ almost all $(x,S) \in \underline{\underline{B}}(0,1) \times \underline{\underline{G}}(n,k)$ and all $t \in [t_2,t_2+R^2]$. Thus

$$(20)\quad \delta(V_t,x_m\psi(t_2+t,x-z))(\underline{\underline{h}}(V_t,x))-\|V_t\|(x_m\partial\psi(t_2+t,x-z)/\partial t)$$

$$\leq M^3[t^{1-3\varepsilon/2}+|z|^{2-2\varepsilon}t^{-\varepsilon/2}+|z|^{4-3\varepsilon}t^{-1}+|z|^{2-\varepsilon}t^{-(1+\varepsilon)/2}]$$

$$+ \underline{\underline{c}}_2M^2[|z|^{2-2\varepsilon}t^{-1/2}+t^{1/2-\varepsilon}]$$

$$+ M^3[t^{1-3\varepsilon/2}+|z|^{2-2\varepsilon}t^{-\varepsilon/2}+|z|^{4-3\varepsilon}t^{-1}+|z|^{2-\varepsilon}t^{-(1-\varepsilon)/2}]$$

$$+ M[t^{1-\varepsilon/2}+|z|^{2-\varepsilon}]t^{-k/2-1}\exp[-1/16t].$$

Integrating t from t_2-q to t_2, we find

$$(21)\qquad \|V_{t_2}\|(x_m\psi(p,x-z))-\|V_{t_2-q}\|(x_m\psi(p+q),x-z))$$

$$\leq 2M^3[(p+q)^{2-3\varepsilon/2}+|z|^{2-2\varepsilon}(p+q)^{1-\varepsilon/2}+|z|^{4-3\varepsilon}|\ell n\ p|+|z|^{2-\varepsilon}(p+q)^{(1-\varepsilon)/2}]$$

$$+ \underline{\underline{c}}_2M^2[|z|^{2-2\varepsilon}(p+q)^{1/2}+(p+q)^{3/2-\varepsilon}]$$

$$+ M[(p+q)^{2-\varepsilon/2}+|z|^{2-\varepsilon}(p+q)](p+q)^{-k/2-1}\exp[-1/16(p+q)].$$

Now take $p+q = R^{\varepsilon}$ and $p = R^3$. Then for $|z| < R$

$$(22) \qquad \|V_{t_2}\|(x_m\psi(p,x-z)) - \|V_{t_2-q}\|(x_m\psi(p+q,x-z))$$

$$\leq 2M^3[R^{3-9\varepsilon/4}+R^{7/2-11\varepsilon/4}+R^{4-3\varepsilon}|\ln R^3|+R^{11/4-7\varepsilon/4}]$$

$$+ \underline{\underline{c}}_2 M^2[R^{11/4-2\varepsilon}+R^{9/4-3\varepsilon/2}]$$

$$+ M[R^{3-3\varepsilon/4}+R^{7/2-\varepsilon}]R^{-3k/4-3/2}\exp[-1/16R^{3/2}]$$

$$< R^{17/8}/3$$

for $\varepsilon = 1/100$ and small enough R.

Next, we show that $\|V_{t_2}\|(x_m\psi(p,x-z))$ is a good approximation to $f_m(t_2,z)$. By judicious rewriting and using the estimates from (19), we find

$$(23) \qquad \|V_{t_2}\|(x_m\psi(p,x-z))$$

$$= \int_T f_m(y)\,\psi(p,x-z)\,|\Lambda_k DF(y)|\,d\mathscr{H}^k y$$

$$= \int_T f_m(z)\,\psi(p,x-z)\,d\mathscr{H}^k y$$

$$+ \int_T (y-z)\cdot Df_m(z)\,\psi(p,x-z)\,d\mathscr{H}^k y$$

$$+ \int_T [f_m(y)-f_m(z)-(y-z)\cdot Df_m(z)]\,\psi(p,x-z)\,d\mathscr{H}^k y$$

$$+ \int_T f_m(y)\,\psi(p,x-z)\,[|\Lambda_k DF(y)|-1]\,d\mathscr{H}^k y$$

$$> f_m(z) + 0 - \int M|y-z|^{2-\varepsilon}\psi(p,x-z)\,d\mathscr{H}^k y$$

$$- \int f_m(y)\psi(p,x-z)k\|S-T\|^2 d\mathscr{H}^k y$$

$$\geq f_m(z) - Mp^{1-\varepsilon/2} - kM^2[p^{2-3\varepsilon/2} + |z|^{2-2\varepsilon}p^{1-\varepsilon/2} + |z|^{4-3\varepsilon} + |z|^{2-}\varepsilon p^{(1-\varepsilon)/2}]$$

$$\geq f_m(z) - MR^3 - kM^2[R^{6-4\varepsilon/2} + R^{5-7\varepsilon/2} + R^{4-3\varepsilon} + R^{7/2-5\varepsilon/2}]$$

$$\geq f_m(z) - R^{17/8}/3$$

for small enough R compared to M.

Next, the smoothness of $\|V_{t_2}\|(x_m\psi(p,x-z))$. Let $H: T \to R$ be defined by

(24) $$H(z) = \|V_{t_2-q}\|(x_m[\psi(p+q,x)$$

$$- z\cdot D\psi(p+q,x) + (z\otimes z/2)\cdot D^2\psi(p+q,x)]).$$

Then by Taylor's theorem with remainder

(25) $$\|V_{t_2-q}\|(x_m\psi(p+q,x-z)) - H(z)$$

$$= \int\int_{\zeta=0}^{1} (1/2)(1-\zeta)^2(z\otimes z\otimes z)\cdot D^3\psi(p+q,x-\zeta z)\,d\zeta\, d\|V_{t_2-q}\|x$$

$$\leq |z|^3\int |D^3\psi(p+q,x)|\,d\|V_{t_2-q}\|x$$

$$\leq |z|^3 M(p+q)^{-3/2}$$

$$\leq R^{3-3\varepsilon/2} < R^{17/8}/3.$$

Thus, combining (22), (23), and (25),

(26) $$f_m(t_2,z) < H(z)+R^{17/8}$$

for $|z| < R$. Looking at things from the opposite direction, we likewise have

(27) $$\tau-f_m(t_2,z) < \|V_{t_2-q}\|(\tau-x_m)[\psi(p+q,x)$$
$$- z\cdot D\psi(p+q,x)+(z\otimes z/2)\cdot D^2\psi(p+q,x)])+R^{17/8}.$$

Hence

(28) $$f_m(t_2,z) > H(z)+\tau-\tau\|V_{t_2-q}\|(\psi(p+q,x))$$
$$+ \tau\|V_{t_2-q}\|[z\cdot D\psi(p+q,x)+(z\otimes z/2)\cdot D^2\psi(p+q,x)].$$

Using the estimates of (19) again,

(29) $$\|V_{t_2-q}\|(\psi(p+q,x)) < \int_T \psi(p+q,x)(1+kDf(t_2,x)^2)\,d\mathscr{H}^k x$$
$$< 1+kM^2(p+q)^{1-\varepsilon},$$

$$\|V_{t_2-q}\|[z\cdot D\psi(p+q,x)] < \int z\cdot D\psi(p+q,x)\,d\mathscr{H}^k x$$
$$+ \int |z|\,|D\psi(p+q,x)|\,kDf(t_2,x)^2 d\mathscr{H}^k x$$
$$< kM^2|z|(p+q)^{1/2-\varepsilon}, \quad \text{and}$$

$$\|V_{t_2-q}\|[(z\otimes z/2)\cdot D^2\psi(p+q,x)] < kM^2|z|^2(p+q)^{-\varepsilon}.$$

Since $\tau < R$, we have for $|z| < R$

(30) $$f_m(t_2,z) > H(z)-2R^{17/8}.$$

By applying Euclidean motions before and after the foregoing analysis, we may find such an H for any direction, any center $y \in \underline{\underline{B}}^k(0,1)$, any time t with $c_{21} < t < t_1-c_{21}$, and any small enough R, i.e. we have $H(t,y,R): T \to T^{\perp}$ such that

(31) $$|f(t,z)-H(t,y,R)(z)| < R^{17/8}$$

for $|y-z| < R$. As in 6.9, we find that $\lim_{R\to 0} H(t,y,R)$ exists, so $f(t,z)$ has second derivatives in z. By comparing $H(t,y_1,R)$ and $H(t,y_2,R)$, we see that $D^2f(t,z)$ is Hölder continuous with exponent $1/8$.

If $Df(t,0) = 0$, then our estimates imply that

$$\begin{aligned} \partial f(t,0)/\partial t &= \lim_{q\to 0} \|V_t\|(T^{\perp}(x)\,\partial\psi(q,x)/\partial t) \\ &= \lim_{q\to 0} \|V_t\|(T^{\perp}(x)\,\Delta\psi(q,x)) \\ &= \Delta f(t,0). \end{aligned}$$

Again, this can be made to apply to any point in $\underline{\underline{B}}^k(0,1)$, so $f(t,z)$ is a classical solution to the non-parametric quasi-linear parabolic partial differential equation discussed in 3.1 for $c_{21} < t < t_1-c_{21}$ and $|z| < 1$. It now follows from standard P.D.E. theory, for example [ES], that $f(t,z)$ is infinitely differentiable.

□

6.11. Local regularity

The smoothness theorem 6.10 requires an absolute bound on the distance of $\operatorname{spt}\|V_t\|$ from a plane T. However, what is available in practice is a bound on the integral of the square of the distance. This lemma links the two.

Lemma: *There are constants* c_{22}, c_{23}, *and* η_0 *such that* $0 < c_{22} < c_{23}$ *and* $\eta_0 > 0$ *and if* V_t *is a unit density integral varifold moving by its mean curvature,*

(1) $$T \in \mathbf{G}(n,k), \quad 0 \le t_0 < \infty, \quad 0 < R < \infty,$$

(2) $$\int_{\mathbf{B}(0,R)} |T^{\perp}(x)|^2 d\|V_{t_0}\|x < \eta_0 R^{k+2}, \quad \text{and}$$

(3) $$\boldsymbol{\alpha}/2 < (R/2)^{-k}\|V_t\|\mathbf{B}(0,R/2) < 3\boldsymbol{\alpha}/2,$$

$$R^{-k}\|V_t\|\mathbf{B}(0,R) < 3\boldsymbol{\alpha}/2,$$

then

$$\{(t,x) \in (t_0+c_{22}R^2, t_0+c_{23}R^2) \times \mathbf{U}(0,R/2) : x \in \operatorname{spt}\|V_t\|\}$$

is an infinitely differentiable manifold.

Proof: We use the clearing out lemma 6.3. We may assume that $T = \mathbf{e}_1 \wedge \cdots \wedge \mathbf{e}_k$, $t_0 = 0$, and $R = 4$.

Suppose $b \in \mathbf{B}^k(0,2)$ and $y \in T^{\perp}+b$ with $|y-b| = 1$. Then $\mathbf{B}(y,1) \subset \mathbf{B}(0,4)$. Define

$$(4)\qquad \phi(x) = \begin{cases} 1-|y-x|^2 & \text{for} \quad |y-x| \le 1 \\ 0 & \text{for} \quad |y-x| \ge 1. \end{cases}$$

Since $1-|y-x|^2 < 2|T^{\perp}(x)|$ for $x \in \underline{\underline{B}}(y,1)$, we have

$$(5)\qquad \|V_0\|(\phi^2) < 4 \int_{\underline{\underline{B}}(y,1)} |T^{\perp}(x)|^2 d\|V_0\|x < 4^{k+3}\eta_0.$$

Hence, by lemma 6.3, for $t > c(2)[4^{k+3}\eta_0]^{2/(k+4)}$

$$(6)\qquad \|V_t\|\underline{\underline{B}}(y,R(t)) = 0 \quad \text{for} \quad R(t)^2 = 1-4kt.$$

Now take τ_0 and c_{21} as in 6.10 and let

$$(7)\qquad r = \tau_0/24kc_{21}, \quad \Delta t = \tau_0^2/432k^2c_{21},$$

and choose η_0 so that

$$(8)\qquad c(2)[4^{k+3}\eta_0]^{2/k+4} = \Delta t.$$

Then for $\Delta t < t < 4\Delta t$, it follows from (6) that for all $x \in \operatorname{spt}\|V_t\| \cap (\underline{\underline{B}}^k(b,r) \times \underline{\underline{B}}^{n-k}(0,1))$,

$$\begin{aligned} y\cdot T^{\perp}(x) &< (1-R(t))+r^2 \\ &< 2k\cdot 4\Delta t+r^2 \\ &< \tau_0^2/54kc_{21} + r\tau_0/24kc_{21} \\ &< \tau_0 r. \end{aligned}$$

Since the direction of y was arbitrary, we actually have $|T^{\perp}(x)| < \tau_0 r$. Also, by (7), we have $\Delta t > c_{21} r^2$. Lemma 6.5 and (3) ensure that 6.10 (4) holds where needed, so theorem 6.10 says that

$$\{(t,x) \in (2\Delta t, 3\Delta t) \times \underline{\underline{U}}^k(k,r/2) \times \underline{\underline{B}}^{n-k}(0,1) : x \in \operatorname{spt}\|V_t\|\}$$

is an infinitely differentiable manifold.

Note that r, Δt, and η_0 are fixed constants, independent of b and V_t, so the above analysis holds for all $b \in \underline{\underline{B}}^k(0,2)$. □

6.12. Main regularity theorem

Finally, we show that V_t is almost everywhere an infinitely differentiable manifold, except when there is a jump decrease in mass.

Theorem: *Suppose V_t is a unit density integral varifold moving by its mean curvature, $t_0 > 0$, $a \in \underline{\underline{R}}^n$, $0 < R_0 < \infty$, and $\lim_{t \uparrow t_0} \|V_t\|(\chi^2(R_0, x-a)) = \|V_{t_0}\|(\chi^2(R_0, x-a))$. Then there is a closed set $B \subset \underline{\underline{R}}^n$ with $\mathscr{H}^k(B) = 0$ such that if $x_0 \in \underline{\underline{B}}(a, R/2) \sim B$, then $\operatorname{spt}\|V_t\|$ is an infinitely differentiable manifold in some neighborhood of (t_0, x_0) in $\underline{\underline{R}}^+ \times \underline{\underline{R}}^n$.*

Remark: Saying "for all $x \notin B$ with $\mathscr{H}^k(B) = 0$" is stronger than saying "for $\|V_{t_0}\|$ almost all x." Indeed, it is possible to have a unit density integral varifold W such that $\operatorname{spt}\|W\|$ is a smooth manifold in some neighborhood of $\|W\|$ almost every point, yet not in any neighborhood of a very large set of points. An example may be constructed by taking W to be an infinite collection of tiny k-spheres that stay away from each other, yet the closure of the set of spheres has positive $\mathscr{H}^n$ measure.

<u>Proof</u>: We may suppose that $R_0 = 1$ and $a = 0$.

If B is the set of points where regularity fails, then B is the complement of a union of open balls, and hence closed.

By the unit density hypothesis and [FH 2.10.19], for $\mathscr{H}^k$ almost all $x \in \underline{\underline{R}}^n$ either $\Theta^k(\|V_{t_0}\|,x) = 1$ and $\operatorname{Tan}^k(\|V_{t_0}\|,x) \in \underline{\underline{G}}(n,k)$, or $\Theta^k(\|V_{t_0}\|,x) = 0$. We shall show that in the first case the local regularity theorem 6.11 can be applied shortly before t_0 except on a set B_1 with $\mathscr{H}^k(B_1) = 0$. In the second case, the clearing out lemma 6.3 will show that a neighborhood of (t_0,x) is empty, except for a set B_2 with $\mathscr{H}^k(B_2) = 0$.

Let B_1 be the set of $x \in \underline{\underline{B}}(0,1/2)$ such that $\Theta^k(\|V_{t_0}\|,x) = 1$ and $\operatorname{Tan}^k(\|V_{t_0}\|,x) \in \underline{\underline{G}}(n,k)$, but $\operatorname{spt}\|V_t\|$ is not a smooth manifold in any neighborhood of (t_0,x). Consider some $b \in B_1$, and let $T = \operatorname{Tan}^k(\|V_{t_0}\|,b)$. Pick $R(b) > 0$ so that

(1) $$\int |T^{\perp}(x)|^2\chi^2(R,x-b)\,d\|V_{t_0}\|x < \eta_0 R^{k+2}2^{-k-3} \quad \text{and}$$

(2) $$\underline{\underline{\beta}}-1/8 < R^{-k} \int \chi^2(R,x-b)\,d\|V_{t_0}\|x < \underline{\underline{\beta}}+1/8$$

for $0 < R < R(b)$. In order for 6.11 not to provide a nice neighborhood of (t_0,b), one of the following must hold for every t and R with $t = t_0 - c_{21}R^2/2$:

(3) $$\int |T^{\perp}(x)|^2\chi^2(R,x-b)\,d\|V_t\|x \geq \eta_0 R^{k+2}2^{-k-2},$$

(4) $$R^{-k} \int \chi^2(R,x-b)\,d\|V_t\|x < \underline{\underline{\beta}}-1/8, \quad \text{or}$$

(5) $$R^{-k} \int \chi^2(R,x-b)\,d\|V_t\|x > \underline{\underline{\beta}}+1/8.$$

Let $B_3(t)$, $B_4(t)$, and $B_5(t)$ be the subsets of B_1 where (3), (4), or (5), respectively, fail at time t.

Define

(6) $$\alpha_b^2(t) = \int_{\underline{\underline{B}}(b,R)} |\underline{\underline{h}}(V_t,x)|^2 d\|V_t\|x \quad \text{and}$$

(7) $$\mu_b^2(t) = \int |T^{\perp}(x)|^2\chi^2(R,x-b)\,d\|V_t\|x.$$

We shall estimate how fast various integrals can change in terms of α and β.

Let $\tilde{\chi} \in \underline{\underline{C}}_0^3(\underline{\underline{R}}^n,\underline{\underline{R}}^+)$ be like χ in depending only on $|x|$, having $\operatorname{spt}\tilde{\chi} \subset \underline{\underline{B}}(0,1)$, and $\tilde{\chi}(x) = 1$ for $x \in \underline{\underline{B}}(0,3/4)$, but also suppose $\tilde{\chi}(x) \leq \chi(x)$ and $|D\tilde{\chi}(x)|$, $\|D^2\tilde{\chi}(x)\| < M\chi(x)$ for all x for some constant M.

Let us look at $B_3(t_1)$ for a particular time t_1. Consider a point $b \in B_3(t_2)$. In going from (3) holding at t_1 to (1) holding at t_0, something must happen to the excess $|T^{\perp}(z)|$. First, we find out what happens if the excess tries moving towards T.

Let R be the above-mentioned radius corresponding to t_1. Define functions $\zeta_b, \zeta_{1b}, \zeta_{2b} \in \underline{C}_0^3(\underline{R}^n, \underline{R}^+)$ by

$$\zeta_{1b}(x) = \tilde{\chi}^2(R,x-b)(1-|T^{\perp}(x-b)|^2/R^2), \tag{8}$$

$$\zeta_{2b}(x) = \chi^2(R,x-b)|T^{\perp}(x-b)|^2/R^2, \quad \text{and}$$

$$\zeta_b(x) = \zeta_{1b}(x) + \zeta_2(x).$$

These functions will be used to detect the motion.

Suppose $t_1 < t < t_0$. Using 3.3 and Schwarz' inequality,

$$\begin{aligned} \delta(V_t,\zeta_1)(\underline{h}(V_t,\cdot)) &= -\int |\underline{h}(V_t,x)|^2 \zeta_1(x)\, d\|V_t\|x \\ &\quad + \int \underline{h}(V_t,x)\cdot S^{\perp}(D\zeta_1(x))\, dV(x,S) \\ &\le \int \underline{h}(V_t,x)\cdot S^{\perp}[2\tilde{\chi}(R,x)D\tilde{\chi}(R,x)(1-|T^{\perp}(x)|^2/R^2) \\ &\quad -2\tilde{\chi}^2(R,x)T^{\perp}(x)/R^2]\, dV_t(x,S) \\ &\le \int 2|\underline{h}(V_t,x)|\,[\tilde{\chi}(R,x)|S^{\perp}(D\tilde{\chi}(R,x))| \\ &\quad +\tilde{\chi}^2(R,x)|T^{\perp}(x)|R^{-2}]\, dV_t(x,S) \end{aligned} \tag{9}$$

$$\leq 4\{\int |\underline{\underline{h}}(V_t,x)|^2\tilde{\chi}^2(R,x)\,d\|V_t\|x$$

$$\cdot [\int |\dot{S}^{\perp}(D\tilde{\chi}(R,x))|^2 + R^{-4}\tilde{\chi}^2(R,x)|T^{\perp}(x)|^2 dV_t(x,S)]\}^{1/2}.$$

From the properties of $\tilde{\chi}$,

$$(10)\quad |S^{\perp}(D\tilde{\chi}(R,x))| \leq |T^{\perp}(D\tilde{\chi}(R,x))| + \|S^{\perp}-T^{\perp}\|\,|D\tilde{\chi}(R,x)|$$

$$\leq MR^{-2}|T^{\perp}(x)|\chi(R,x) + \|S-T\|\,|D\tilde{\chi}(R,x)|,$$

so, using 5.4 with $\phi = |D\tilde{\chi}(R,x)|$,

$$(11)\qquad \int |\dot{S}^{\perp}(D\tilde{\chi}(R,x))|^2 dV_t(x,S)$$

$$\leq 2M^2R^{-4}\int |T^{\perp}(x)|^2\chi^2(R,x)\,d\|V_t\|x$$

$$+ 2\int \|S-T\|^2\, D\tilde{\chi}(R,x)|^2 dV_t(x,S)$$

$$\leq 2M^2R^{-4}\mu_b^2(t) + 32\int |D|D\tilde{\chi}(R,x)||^2|T^{\perp}(x)|^2 d\|V_t\|x$$

$$+ 4[\int |\underline{\underline{h}}(V_t,x)|^2|D\tilde{\chi}(R,x)|^2 d\|V_t\|x \int |D\tilde{\chi}(R,x)|^2|\dot{T}^{\perp}(x)|^2 d\|V_t\|x]^{1/2}$$

$$\leq 2M^2R^{-4}\mu_0^2(t) + 32M^2R^{-4}\mu_b^2(t) + 4M^2R^{-2}\alpha_b(t)\mu_0(t)$$

$$\leq 34M^2R^{-4}\mu_b^2(t) + 4M^2R^{-2}\alpha_b(t)\mu_b(t).$$

Thus

$$(12)\quad \delta(V_t,\zeta_1)(\underline{\underline{h}}(V_t,\cdot)) \leq 24MR^{-2}\alpha_b(t)\mu_b(t) + 8MR^{-1}\alpha_b(t)^{3/2}\mu_b(t)^{1/2}.$$

Next, we make for imminent use the estimate

(14) $$\int |\underline{\underline{h}}(V_t,x)|\,|D\zeta_2(x)|\,d\|V_t\|x$$

$$\leq \int |\underline{\underline{h}}(V_t,x)|2R^{-2}[\chi(R,x)|D\chi(x/R)|\,|T^{\perp}(x)|^2$$

$$+ \chi^2(R,x)|T^{\perp}(x)|]d\|V_t\|x$$

$$\leq 2\{\int |\underline{\underline{h}}(V_t,x)|^2|D\chi(R,x)|^2|T^{\perp}(x)|^2R^{-2}d\|V_t\|x$$

$$\cdot\int \chi^2(R,x)|T^{\perp}(x)|^2d\|V_t\|x\}$$

$$+ 2R^{-2}\{\int |\underline{\underline{h}}(V_t,x)|^2\chi^2(R,x)d\|V_t\|x$$

$$\cdot\int \chi^2(R,x)|T^{\perp}(x)|^2d\|V_t\|x\}$$

$$\leq 4MR^{-2}\alpha_b(t)\mu_b(t).$$

Now suppose $\mathscr{B}$ is a collection of disjoint balls $\underline{\underline{B}}(b,R)$ with $b \in B_3(t_1)$. By 3.3, for $t_1 < t < t_0$,

(15) $$\delta(V_t,\chi^2-\sum_{\mathscr{B}}\zeta_b)(\underline{\underline{h}}(V_t,\cdot))$$

$$\leq -\int |\underline{\underline{h}}(V_t,x)|^2(\chi^2(x)-\sum_{\mathscr{B}}\zeta_b(x))d\|V_t\|x$$

$$+ \int |\underline{\underline{h}}(V_t,x)|2\chi(x)|D\chi(\alpha)|d\|V_t\|x$$

$$+ \sum_{\mathscr{B}}\int |\underline{\underline{h}}(V_t,x)|\,|S^{\perp}(D\zeta_1(x))|dV_t(x,S)$$

$$+ \sum_{\mathscr{B}}\int |\underline{\underline{h}}(V_t,x)|\,|S(D\zeta_2(x))|dV_t(x,S).$$

By Minkowski's inequality,

$$\int |\underline{\underline{h}}(V_t,x)|2\chi(x)|D\chi(x)|d\|V_t\|x$$

$$\leq \int_{\operatorname{spt} D\chi} |\underline{\underline{h}}(V_t,x)|^2\chi^2(x)d\|V_t\|x + \int |D\chi(x)|^2 d\|V_t\|x.$$

Since $\operatorname{spt} \zeta_b \cap \operatorname{spt} D\chi = \phi$ for each b in the sum, the first two terms on the right hand side of (15) are dominated by $\int |D\chi(x)|^2 d\|V_t\|x$. The third term is what was actually estimated in (9)-(12), and the fourth term is taken care of by (14). Thus (15) becomes

$$(16) \qquad \delta(V_t,\chi^2-\sum_{\mathcal{B}}\zeta_b)(\underline{\underline{h}}(V_t,\cdot)) \leq \int |D\chi(x)|^2 d\|V_t\|x$$

$$\sum_{\mathcal{B}}[28MR^{-2}\alpha_b(t)\mu_b(t)+8MR^{-1}\alpha_b(t)^{3/2}\mu_0(t)^{1/2}].$$

From (12) and (16) we may calculate that

$$(17) \qquad \|V_{t_0}\|(\chi^2)-\|V_{t_*}\|(\chi^2) \leq \|V_{t_0}\|(\sum_{\mathcal{B}}\delta_{1b})-\|V_{t_*}\|(\sum_{\mathcal{B}}\delta_{1b})$$

$$+ \int_{t_*}^{t_0} \delta(V_t,\sum_{\mathcal{B}}\zeta_{2b})(\underline{\underline{h}}(V_t,\cdot))dt$$

$$+ \int_{t_*}^{t_0} \delta(V_t,\chi^2-\sum_{\mathcal{B}}\zeta_b)(\underline{\underline{h}}(V_t,\cdot))dt$$

$$\leq R^{-2}\sum_{\mathcal{B}}[\mu_b^2(t_0)-\mu_b^2(t_*)] + \int_{t_*}^{t_0} \|V_t\|(|D\chi|^2)dt$$

$$+ \sum_{\mathcal{B}} \int_{t_*}^{t_0} [52MR^{-2}\alpha_b(t)\mu_b(t)+16MR^{-1}\alpha_b(t)^{3/2}\mu_0(t)^{1/2}]dt.$$

Now we choose $\mathscr{B}$ and t_* to use in the above. Let $B_3^*(t_1)$ be the set of points $b \in B_3(t_2)$ such that

(18) $$\|V_{t_0}\|\underline{\underline{B}}(b,R) < (4/3)\underline{\underline{\alpha}}R^k.$$

If $\limsup_{t_1 \uparrow t_0} \mathscr{H}^k(B_3(t_1)) > 0$, then the unit density hypothesis implies that there are t_1 arbitrarily close to t_0 such that

(19) $$\mathscr{H}^k(B_3^*(t_1)) > (3/4)\mathscr{H}^k(B_3(t_1)).$$

By the Besicovitch covering theorem 2.2, we can find a disjoint collection $\mathscr{B}$ of balls $\underline{\underline{B}}(b,R)$ with $b \in B_3(t_1)$ and

(20) $$\|V_{t_0}\|(\cup\mathscr{B}) \geq B(n)^{-1}\|V_{t_0}\|B_3^*(t_1).$$

We infer from (18), (19), (20), and unit density that

(21) $$\begin{aligned} \sum_{\mathscr{B}}\underline{\underline{\alpha}}R^k &> (3/4)\sum_{\mathscr{B}}\|V_{t_0}\|\underline{\underline{B}}(b,R) \\ &> (3/4\underline{\underline{B}}(n))\|V_{t_0}\|B_3^*(t_1) \\ &> (3/4\underline{\underline{B}}(n))\mathscr{H}^k(B_3^*(t_1)) \\ &> (1/2\underline{\underline{B}}(n))\mathscr{H}^k(B_3(t_1)). \end{aligned}$$

Now let t_* be a value of t for which $\sum_{\mathscr{B}}\mu_b^2(t)$ is nearly maximal for $t_1 \leq t \leq t_0$. From (1), (3), and (21) we see that we may assume that

(22) $$\sum_{\mathscr{B}}[\mu_b^2(t_*)-\mu_0^2(t_0)] \geq (1/2R^2)\sum_{\mathscr{B}}\mu_b^2(t_*)$$

$$\geq \sum_{\mathscr{B}}\eta_0 R^k 2^{-k-3}$$

$$\geq 2^{-k-4}\underline{\underline{B}}(n)^{-1}\eta_0\mathscr{H}^k(B_3(t_1)).$$

Note that this last estimate does not depend explicitly on R.

Suppose that

$$\limsup_{t_1\uparrow t_0} \mathscr{H}^k(B_3(t_1)) > 0.$$

By hypothesis, $\lim_{t\uparrow t_0}\|V_t\|(\chi^2) = \|V_{t_0}\|(\chi^2)$, and barrier functions may be used to show that $\|V_t\|(|D\chi|^2)$ is bounded. Thus (22) shows that we may assume that t_1 is such that (17) implies

$$R^{-2}\sum_{\mathscr{B}}[\mu_b^2(t_*)-\mu_b^2(t_0)]$$

$$\leq 100M\sum_{\mathscr{B}}\int_{t_*}^{t_0}[R^{-2}\alpha_b(t)\mu_b(t)+R^{-1}\alpha_b(t)^{3/2}\mu_0(t)^{1/2}]dt$$

$$\leq 100MR^{-2}\sum_{\mathscr{B}}2\mu_b(t_*)(t_0-t_*)^{1/2}\{\int_{t_*}^{t_0}\alpha_b^2(t)\,dt\}^{1/2}$$

$$+\ 100MR^{-1}\sum_{\mathscr{B}}2\mu_b(t_*)^{1/2}(t_0-t_*)^{1/4}\{\int_{t_*}^{t_0}\alpha_b^2(t)\,dt\}^{3/4}.$$

Using (22), $t_0 - t_* < c_{21}R^2$, and Schwarz' inequality on the sums, we get

$$(1/400M)\sum_{\mathcal{b}}\mu_b^2(t_*)$$

$$\leq c_{21}^{1/2}R^{-1}\{\sum\mu_b^2(t_*)\}^{1/2}\{\sum_{\mathcal{b}}\int_{t_*}^{t_0}\alpha_b^2(t)\,dt\}^{1/2}$$

$$+ c_{21}^{1/4}R^{-1/2}\{\sum_{\mathcal{b}}\mu_b^2(t_*)\}^{1/4}\{\sum_{\mathcal{b}}\int_{t_*}^{t_0}\alpha_b^2(t)\,dt\}^{3/4}.$$

This implies that

$$(23)\qquad \sum_{\mathcal{b}}\int_{t_*}^{t_0}\alpha_b^2(t)\,dt \geq \sum_b\mu_b^2(t_*)/400^2M^2c_{21}R^2.$$

Thus, using (22),

$$(24)\qquad \sum_{\mathcal{b}}\int_{t_*}^{t_0}\alpha_b^2(t)\,dt > 2^{-k-22}\underline{\underline{B}}(n)^{-1}M^{-2}c_{21}^{-1}\eta_0\mathscr{H}^k(B_3(t_1)).$$

But now we can easily calculate from 3.3 and Minkowski's inequality that

$$(25)\qquad \|V_{t_0}\|(\chi^2)-\|V_{t_*}\|(\chi^2) \leq \int_{t_*}^{t_0}\delta(V_t,\psi)(\underline{h}(V_t,\cdot))\,dt$$

$$\leq -(1/2)\int_{t_*}^{t_0}\int|\underline{h}(V_t,x)|^2\chi^2(x)\,d\|V_t\|x\,dt$$

$$+ 4\int_{t_*}^{t_0}\|V_t\|(|D\chi(x)|^2)\,dt.$$

We know $\|V_t\|(|D\chi(x)|^2)$ is bounded, so (24) implies

$$(26)\qquad \|V_{t_*}\|(\chi^2) \leq \|V_{t_0}\|(\chi^2)-2^{-k-23}\underline{\underline{B}}(n)^{-1}M^{-2}c_{21}^{-1}\eta_0\mathscr{H}^k(B_3(R)).$$

By hypothesis, $\lim_{t \to t_0} \|V_t\|(\chi^2) = \|V_{t_0}\|(\chi^2)$, so we finally have

(27) $$\limsup_{t_1 \uparrow t_0} \mathscr{H}^k(B_3(t_1)) = 0.$$

If $B_4(t_1)$ and $B_5(t_1)$ are the subsets of B_1 where (4) or (5) fail respectively, then similar (but simpler) analyses show that

(28) $$\limsup_{t_1 \to t_0} \mathscr{H}^k(B_4(t_1)) = 0 \quad \text{and}$$

(29) $$\limsup_{t_1 \uparrow t_0} \mathscr{H}^k(B_5(t_1)) = 0.$$

Since $B_1 = B_3(t_1) \cup B_4(t_1) \cup B_5(t_1)$ for any t_1, clearly we must have $\mathscr{H}^k(B_1) = 0$.

For the second part of the proof, let B_2 be the set of points $x \in \underline{B}(0,1/2)$ with $\Theta^k(\|V_{t_0}\|,x) = 0$, but for every $R > 0$ having $\|V_t\|\underline{B}(x,R) > 0$ for some $t \in [t_0-R^2, t_0+R^2]$. Let $c(3)$ be as in the clearing out lemma 6.3, and define

(30) $$\phi(x) = \begin{cases} 1-|x|^2 & \text{for} \quad |x| \leq 1, \\ 0 & \text{for} \quad |x| \geq 1. \end{cases}$$

Choose $\eta > 0$ so that

(31) $$4kc(3)\eta^{2/(k+6)} < 1/2.$$

Let $B_6(R)$ be the set of $b \in B_2$ such that

(32) $$\int \phi^3((x-b)/R)\, d\|V_{t_0}\|x < (\eta/2)R^k.$$

Let $t_1 = t_0 - R^2/6k$. Then the clearing out lemma implies that for $b \in B_2$

(33) $$\int \phi^3((x-b)/R)\, d\|V_{t_1}\|x \geq \eta R^k.$$

It follows from the definition of Hausdorff measure and the Besicovitch covering theorem that there is a collection $\mathcal{B}$ of disjoint balls $\underline{\underline{B}}(b,R)$ with $b \in B_b(R)$ and

(34) $$\sum_b \underline{\underline{\alpha}} R^k > (1/2\underline{\underline{B}}(n))\mathcal{H}^k(B_6(R)).$$

Equations (32) and (33) show that mass is being lost from $\cup\mathcal{B}$, and we next show that it can't be going elsewhere. We estimate for $t_1 < t < t_0$:

(35) $$\delta(V_t, \chi^2 - \sum_b \phi^3((\cdot - b)/R))(\underline{\underline{h}}(V,\cdot))$$

$$\leq -\int |\underline{\underline{h}}(V_t,x)|^2(\chi^2(x) - \sum_b \phi^3((x-b)/R))\, d\|V_t\|x$$

$$+\int |\underline{\underline{h}}(V_t,x)|2\chi(x)|D\chi(x)|\, d\|V_t\|x$$

$$+\sum_b \int |\underline{\underline{h}}(V_t,x)|3\phi^2((x-b)/R)|D\phi((x-b)/R)|\, d\|V_t\|x$$

$$\leq \int |D\chi(x)|^2 d\|V_t\|x$$

$$+\ 3\sum_b\{\int|\underline{\underline{h}}(V_t,x)|^2|D\phi((x-b)/R)|^2\phi((x-b)/R)\,d\|V_t\|x$$

$$\cdot\int\phi^3((x-b)/R)\,d\|V_t\|x\}^{1/2}.$$

Define

$$(36)\qquad \alpha_b^2(t) = \int_{\underline{\underline{B}}(b,R)} |\underline{\underline{h}}(V,x)|^2 d\|V_t\|x,$$

$$(37)\qquad \beta_b^2(t) = \int \phi^3((x-b)/R)\,d\|V_t\|x.$$

Thus

$$(38)\qquad \delta(V_t,\psi-\sum_k\phi^3(\cdot-b)/R))(\underline{\underline{h}}(V,\cdot))$$

$$\leq \|V_t\|(|D\chi|^2)+6R^{-1}\alpha_b(t)\beta_b(t)$$

Let t_* be a time for which $\sum_b \beta_b^2(t)$ has nearly its maximum value between t_1 and t_0. Integrating this from t_1 to t_0 and using (32) and (33) we get

$$(39)\qquad \|V_{t_0}\|(\psi)-\|V_{t_*}\|(\psi) \leq \sum_b \beta_b^2(t_0)-\beta_0^2(t_*)$$

$$+\int_{t_*}^{t_0}\|V_t\|(|D\chi|^2)\,dt+6R^{-1}\sum_b\int_{t_*}^{t_0}\alpha_b(t)\beta_b(t)\,dt.$$

Again we may neglect the terms explicitly containing χ and $D\chi$. By Minkowski's inequality, (32), (33) and the definitions of t_*, (39) then becomes

$$(40)\qquad \{\sum_b \beta_b(t_*)^2(t_0-t_*)\}^{1/2}\{\sum_\infty \int_{t_*}^{t_0} \alpha_b^2(t)\,dt\}^{1/2}$$

$$> \sum_b (\beta_b^2(t_*)-\beta_b^2(t_0))R/6$$

$$> (R/12)\sum_b \beta_b^2(t_*).$$

Hence

$$(41)\qquad \sum_b \int_{t_*}^{t_0} \alpha_b^2(t)\,dt > (1/12)\sum_b \beta_b^2(t_*)$$

$$> (\eta/24\underline{\underline{B}}(n)\underline{\underline{\alpha}}\mathscr{H}^k(B_6(R)).$$

As before, this implies

$$\limsup_{R\to 0} B_6(R) = 0.$$

Since very $b \in B_2$ is in $B_6(R)$ for small enough R, this implies $\mathscr{H}^k(B_2) = 0$. □

Appendix A: Grain growth in metals

The purposes of this appendix are to describe a physical system involving motion by mean curvature and to correct a calculation made in [RCD].

The physical system is the motion of grain boundaries in an annealing piece of metal such as aluminum. The lowest energy state of aluminum at a temperature just below its melting point is a certain crystalline lattice. However, when a sample of molten aluminum solidifies, crystallization may start in many different places with random orientations, and the solid metal will be composed of many small regions, each with uniform crystal structure. Each connected such region is called a grain.

An atom on a grain boundary is only partially surrounded by a nice lattice; therefore it is in a higher energy state than an atom in the interior of a grain. This extra energy may be thought of as endowing the grain boundary with a surface tension. The size of this surface tension should be about the same order of magnitude as the surface tension of the liquid metal [CH]. The surface tension of aluminum at its melting point is 860 ergs/cm^2 [HCP p. F-19] (which may be compared to that of water at 18°C, which is 73 ergs/cm^2 [HCP p. F-33]). It would be expected that the surface tension of a grain boundary would depend on the orientations of the grains bounded. However, experimentally the dependence seems small except for small differences in orientation [SC1].

It is observed that if pure aluminum with many small grains is annealed, then the grain boundaries move with velocities proportional to their mean curvatures [RCD]. On an atomic scale, the motion may be viewed as due to the probability of an atom at a grain boundary finding itself, as a result of random thermal motion, on one or another of the adjacent lattices. Clearly the probability of landing in a concave lattice is greater than landing in a convex one, and the measure of the difference of probabilities in general is the mean curvature [SC2].

Assuming the surface tension independent of orientation, then by [TJ] one should find throughout the motion that three boundaries meet at 120° angles in a line and four boundaries meet at approximately 109° angles at a point. In one sample of aluminum, over 3000 junctions were examined without finding any other configurations [RCD].

Since an arrangement of grains such that the boundaries have no mean curvature, for example, a stacking of Kelvin's tetrakaidecahedron [K], is extremely unlikely, boundary motion continues until the sample consists of a few large grains. Relatively larger grains tend to have more faces than smaller grains and thus the average face on a large grain tends to be more concave. Therefore large grains grow at the expense of small ones. The dividing line between growing and shrinking seems to be at about 14 faces [SC2]. The general distribution of shapes seems to be independent of average grain size [RCD].

The assumption that the distribution of grain shapes is independent of time enables one to estimate the rate of growth as a function of several physical constants. This calculation was unsuccessfully attempted in [RCD]. Suppose we start at time $t = 0$ with a sample whose average grain size is assumed ideally to be zero. Define

γ = surface tension of boundary,

$S(t)$ = average boundary area per unit volume,

$N(t)$ = average number of grains per unit volume,

$H(t)$ = average magnitude of the mean curvature

μ = mobility of the boundary; i.e., the velocity is mobility times pressure,

σ = $H(t)S^2(t)/N(t)$,

k = $S(t)N(t)^{-1/3}$,

Θ = ratio of the volume of the average grain to the volume swept out by the boundary during the disappearance of an average grain.

Note that σ, k, and Θ are dimensionless and thus constant in time because of the assumption of the constancy of the distribution of grain shapes. The average pressure on the boundaries is

$$P(t) = \gamma H(t).$$

The rate at which volume is being swept out per unit volume is

$$\mu P(t) S(t) = \mu\gamma H(t) S(t).$$

The rate of grain loss per unit volume is

$$dN(t)/dt = -\mu\gamma H(t)S(t)\Theta N(t).$$

Using the constants σ and k to eliminate $H(t)$ and $S(t)$ we get

(1) $$dN(t)/dt = -\mu\gamma\Theta\sigma k^{-1}N(t)^{-5/3}$$

which has the solution

$$N(t) = ((2/3)\mu\gamma\Theta\sigma k^{-2}t)^{-3/2}$$

which gives the expected power law dependence of volume on time [SC2]. We can derive the value of Θ from the principle of conservation of energy: the work done against "friction" by the moving boundaries must be less than or equal to the energy released by the shrinking of the area of the boundaries. By the work done against "friction" I mean force times speed, or pressure times area times speed. Thus

$$\mu\gamma^2H(t)^2S(t) \leq -\gamma dS(t)/dt.$$

Using σ and k again, we have

$$\mu\gamma^2\sigma^2k^{-3}N(t) \leq -\gamma k dN(t)^{1/3}/dt,$$

and using (1) for $dN(t)/dt$ yields

$$\Theta^{-1} \leq k^3/3\sigma.$$

From experimental data presented in [RCD fig. 4,11] we get $\sigma \approx 1.33$ and $k^3 \approx 10$, so $\Theta^{-1} \lesssim 2.5$. It seems entirely reasonable for boundaries to sweep out 2.5 grain volumes during a disappearance.

Appendix B: Curves in R^2

The simplest non-trivial class of varifolds moving by mean curvature is the class of smooth closed curves in $\underline{\underline{R}}^2$. Even here, exact solutions are hard to find, so we will be content with deriving some general properties which give a feeling for the effects of motion by mean curvature.

Because we are dealing only with times when a curve is smooth, we will use the mapping approach discussed in 3.1. Suppose $t_1 \in \underline{\underline{R}}^+$,

$$F: [0,t_1] \times \underline{\underline{S}}^1 \to \underline{\underline{R}}^2$$

is smooth, and $F(t,\cdot)$ is non-self-intersecting closed curve for each $t \in [0,t_1]$.

Define the metric $g: [0,t_1] \times \underline{\underline{S}}^1 \to \underline{\underline{R}}^+$ by

$$g(t,\theta) = |\partial F(t,\theta)/\partial\theta|,$$

the tangent angle $\beta: [0,t_1] \times \underline{\underline{S}}^1 \to \underline{\underline{R}}$ by

$$\tan \beta(t,\theta) = (\underline{\underline{e}}_2 \cdot \partial F(t,\theta)/\partial\theta)/(\underline{\underline{e}}_1 \cdot \partial F(t,\theta)/\partial\theta)$$

and the oriented curvature $K: [0,t_1] \times \underline{\underline{S}}^1 \to \underline{\underline{R}}$ by

$$K(t,\theta) = g(t,\theta)^{-1}\partial\beta(t,\theta)/\partial\theta.$$

If $F(t,\cdot)$ is moving by its mean curvature in the mapping sense, then it can be derived that for all $(t,\theta) \in [0,t_1] \times \underline{S}^1$

(1) $$\partial g(t,\theta)/\partial t = -g(t,\theta)K(t,\theta)^2,$$

(2) $$\partial K(t,\theta)/\partial t = g(t,\theta)^{-2}\partial^2 K(t,\theta)/\partial\theta^2 + K(t,\theta)^3.$$

Suppose $K(t,\theta)$ is positive when the mean curvature vector points toward the inside of the curve.

PROPOSITION 1: *The area enclosed by the curve* $F(t,\cdot)$ *decreases at the rate of* 2π *for all* $t \in [0,t_1]$.

Proof: The rate at which area $A(t)$ decreases is given by

$$dA(t)/dt = -\int K(t,\theta)g(t,\theta)\,d\theta$$

which is -2π by the Gauss Bonnet theorem. □

PROPOSITION 2: *The total curvature of* $F(t,\cdot)$ *is monotone decreasing for each* $t \in [0,t_1]$.

Proof: The rate of change of total curvature is

$$(\partial/\partial t)\int |K(t,\theta)|g(t,\theta)\,d\theta$$

$$= \int [\text{sign } K(t,\theta)](\partial K(t,\theta)/dt)g(t,\theta)$$

$$+ |K(t,\theta)|(\partial g(t,\theta)/\partial t)\,d\theta,$$

using (1) and (2)

$$= \int [\text{sign } K(t,\theta)][g(t,\theta)^{-2}\partial^2 K(t,\theta)/\partial\theta^2 + K(t,\theta)^3]$$

$$\cdot g(t,\theta) - [\text{sign } K(t,\theta)]K(t,\theta)^3 d\theta$$

$$= \int [\text{sign } K(t,\theta)][\partial^2 K(t,\theta)/\partial\theta^2] g(t,\theta)^{-1} d\theta$$

$$\leq 0.$$

The last inequality follows by integrating over intervals of θ where $K(t,\theta)$ has constant sign: Suppose $a < b$, $K(t,a) = K(t,b) = 0$, and $K(t,\theta) > 0$ for $a < \theta < b$. Let $\sigma(\theta)$ be arc-length, so $d\sigma = g(t,\theta)d\theta$. Then

$$\int_a^b [\partial^2 K(t,\theta)/\partial\theta^2] g(t,\theta)^{-1} d\theta$$

$$= \int_{\sigma(a)}^{\sigma(b)} (\partial^2 K(t,\theta(\sigma))/\partial\sigma^2) d\sigma$$

$$= \partial K(t,b)/\sigma \;-\; \partial K(t,a)/\partial\sigma.$$

By hypothesis, $\partial K(t,b)/\partial\sigma \leq 0$ and $\partial K(t,a)/\partial\sigma \geq 0$, so we are done. A similar analysis can be done for intervals on which $K < 0$. □

Definition: If F_1, F_2: $[0,t_1] \times \underline{\underline{S}}^1 \to \underline{\underline{R}}^2$ are two smooth curves moving by their mean curvature, then define for each $t \in [0,t_1]$

the *area between* $F_1(t,\cdot)$ *and* $F_2(t,\cdot)$, denoted $B(t)$, by

$$B(t) = \mathcal{L}^2\{x \in R^2 : x \text{ inside exactly one of } F_1(t,\cdot) \text{ or } F_2(t,\cdot)\}.$$

Definition: We say that the *orders of mutual intersections* of two curves are the same if the two curves intersect at a finite number of points and the orders of these intersections are the same along both curves.

PROPOSITION 3: *If* F_1 *and* F_2 *are two smooth curves moving by their mean curvatures, and if the orders of mutual intersections are the same along both curves for each* t, *then the total area between* $F_1(t,\cdot)$ *and* $F_2(t,\cdot)$ *is monotonically decreasing for each* t.

Proof: If $F_1(t,\cdot)$ and $F_2(t,\cdot)$ do not intersect at all for a particular t, then it follows from Proposition 1 that $B(t)$ is constant until the curves intersect or one curve vanishes.

Otherwise, suppose without loss of generality that $a, b \in \underset{\sim}{S}^1$ are parameters of successive intersections as indicated in figure 1. Then the rate of change of the area $B(t)$ of the region between a and b is

$$dB(t)/dt = \int_a^b K_1(t,\theta) g_1(t,\theta)\, d\theta - \int_a^b K_2(t,\theta) g_2(t,\theta)\, d\theta$$

$$= \beta_1(t,b) - \beta_1(t,a) - \beta_2(t,b) + \beta_2(t,a)$$

$$= [\beta_1(t,b) - \beta_2(t,b)] + [\beta_2(t,a) - \beta_1(t,a)]$$

$$\leq 0. \qquad \Box$$

These three propositions suggest that in general dimensions the "area bounded" ought to decrease, the total curvature ought to decrease, and two surfaces starting out nearly alike should get even more alike.

Proposition 1 depends on the mean curvature being the same as the Gaussian curvature, which is not true in higher dimensions. For example, if the original surface is a 2-sphere in $\underline{\underline{R}}^3$ with a lot of sharp inward spikes, then the volume enclosed will increase at first.

It seems intuitively clear that a surface will locally smooth itself out and thus reduce its total mean curvature. However, global effects may reverse this. Figure 2 illustrates how this might happen. The 2-surface in $\underline{\underline{R}}^3$ is two infinite parallel flat sheets with a large diameter doughnut hole. Since the curvature due to the closeness of the sheets is greater than that due to the diameter of the hole, the hole will expand. Therefore the area of the region with high curvature expands, so the total curvature increases.

Proposition 3 implies a continuous dependence on initial conditions if nearness is measured in terms of area between curves. While continuous dependence on initial conditions would hold for higher dimensional smooth manifolds, it does not hold for general surfaces. For example, let the initial surface be two unit circles in $\underline{\underline{R}}^2$ distance d apart, $d \geq 0$. If $d > 0$, the only possible course is for the circles to shrink down to their respective centers. If $d = 0$ then the circles may remain connected and turn into a dumbbell shape.

Appendix C: Curves of constant shape

C.1. In this appendix we investigate one-dimensional integral varifolds V_t in $\underline{\underline{R}}^2$ moving by their mean curvature such that if $s, t > 0$ then V_t is a homothety of V_s. Suppose we have such a varifold V_t and $R: \underline{\underline{R}}^+ \to \underline{\underline{R}}^+$ is such that for $t > 0$

$$V_t = \underline{\underline{\mu}}(R(t))_\# V_1.$$

Then $R(t)$ is a characteristic scale of V_t, and since mean curvature is inversely proportional to scale, we must have

$$dR(t)/dt = \beta/R(t)$$

for some $\beta \in \underline{\underline{R}}$. Then

$$R(t) = (2\beta t + R(0)^2)^{1/2}.$$

Note that this scaling factor is valid for all dimensions.

We now seek a differential equation describing curves in $\underline{\underline{R}}^2$ which remain homotheties of themselves. Let the curve be given by $F: \underline{\underline{R}}^+ \times \underline{\underline{R}} \to \underline{\underline{R}}^2$ with scaling factor $R(t)$. For a particular time $t \in \underline{\underline{R}}^+$, let $\gamma = dR(t)/dt$, $\underline{\underline{n}}: \underline{\underline{R}} \to \underline{\underline{R}}^2$ be the normal vector, $\underline{\underline{h}}: \underline{\underline{R}} \to \underline{\underline{R}}^2$ the mean curvature vector, and $K: \underline{\underline{R}} \to \underline{\underline{R}}$ be the oriented scalar curvature of $F(t,\cdot)$ so that $K(s)\underline{\underline{n}}(s) = \underline{\underline{h}}(s)$ for all $t \in \underline{\underline{R}}$. The condition we are looking for is, for all $s \in \underline{\underline{R}}$

(1) $$\gamma F(t,s)\cdot \underline{\underline{n}}(s) = K(s).$$

Given γ, $F(t,0)$, and the initial direction of $F(t,\cdot)$, one can integrate (1) to get a curve. This I have done on an HP 9820A desktop calculator with plotter, and some of the results are described in the following sections.

C.2. Corners

If one takes $\gamma = 2$, $F(t,0) = (0,3/8)$ and initial direction horizontal, one gets a curve as in Figure 3, which has asymptotes at approximately a right angle. If the time origin is chosen so that $R(0) = 0$, then we see that the initial surface was approximately a right angle. Thus we know what the evolution of a corner looks like.

C.3. Triple junctions

We take $\gamma = 1$ and three curves starting at $(.43,0)$ at angles of 0°, 120°, and 240°. Since the triple junction contributes no curvature, and everywhere else obeys (1), this represents a curve of constant shape. See Figure 4. The parameters were chosen so that the initial surface was a vertical line with a horizontal line meeting it at the origin.

C.4. Multiple rays

Figure 5 shows a possible evolution of two lines meeting at right angles. The parameters are $\gamma = 1$, starting points

(0,.28) and (0,-.28), and starting angles 120° apart.

Note the essential non-uniqueness of the evolution. By symmetry, the same solution rotated 90° is also a solution.

This solution will arise from the reduced mass model (see 4.9 Remark 2) but not in the normal varifold model. Since the initial varifold has zero mean curvature, one must ask what Lipschitz maps of small displacement can do. If one is not allowed to reduce mass, then any non-identity Lipschitz map will increase mass. Thus in the normal model nothing will happen. In the reduced mass model, the first small Lipschitz map can produce a miniature version of Figure 5, and the rest of the evolution is driven by mean curvature.

Figures 6, 7, and 8 show possible evolutions of 5, 7, and 8 rayed initial surfaces with no special angles.

C.5. Shrinking loop

If one considers $\beta < 0$, then one gets surfaces that shrink as time increases. A circle is the most obvious example. Another interesting example is the loop shown in Figure 9. Its parameters are $\gamma = -1$, starting point (0,.83), and angles 90°, 210°, and 330°. The starting point was chosen so the two lower curves joined smoothly below the origin. This surface will shrink homothetically until the loop vanishes, leaving a vertical ray from the origin. The ray will then vanish instantaneously. This is an example of non-continuity, and thus non-differentiability.

Appendix D: Density bounds and rectifiability

The example presented in this appendix illustrates the necessity of assuming a lower bound on the density of a rectifiable initial varifold in 4.1 in order to conclude rectifiability later in 4.17. We construct an initial rectifiable varifold with lower density bound zero and give an argument that this varifold should turn unrectifiable as it moves by its mean curvature.

First, let W be the varifold depicted in Figure 10. The densities are to be such that W is stationary. We intend that $W = 2\underline{\underline{\mu}}(2)_{\#}W$.

Define $\beta: \underline{\underline{R}}^2 \times \underline{\underline{G}}(2,1) \to \underline{\underline{R}}^+$ by

$$\beta(x_1, x_2, S) = \exp|x_2|.$$

Let $V_0 = W \llcorner \beta \in \underline{\underline{RV}}_1(\underline{\underline{R}}^2)$. Note that β was chosen to give V_0 unit magnitude mean curvature vectors on the vertical segments pointing away from the horizontal centerline.

To see that this initial varifold will evolve as claimed under the construction of Chapter 4, consider the mth approximation. Away from the centerline, motion is outward with a more or less uniform velocity, which preserves the density gradients, which preserves the uniform outward velocity. The stuff near the centerline will be vertically stretched by the small Lipschitz maps f_2 (see 4.9). When we take the limit of the approximations as $m \to \infty$, the stretched central stuff converges to a region of one-dimensional varifold expanding with unit speed with zero one-dimensional density but with positive two-dimensional density. It is thus unrectifiable.

FIGURE CAPTIONS

FIGURE 1. Portions of two moving smooth closed curves and the shrinking area between them.

FIGURE 2. A doughnut hole, which has increasing total mean curvature as it evolves.

FIGURE 3. A stage in the evolution of an initial right angle. All stages have a mathematically similar shape.

FIGURE 4. Evolution of three lines meeting at right angles.

FIGURE 5. Evolution of four lines meeting at right angles.

FIGURE 6. Evolution of five lines meeting at random angles.

FIGURE 7. Evolution of seven lines.

FIGURE 8. Evolution of eight lines.

FIGURE 9. A one dimensional surface which evolves by the loop shrinking down to a point, leaving a line that vanishes instantaneously.

FIGURE 10. A rectifiable one dimensional initial varifold which intuitively should evolve into an unrectifiable varifold. The pattern continues indefinitely towards the center with decreasing line weights. Along the center line, one dimensional densities are zero, but two dimensional densities are not.

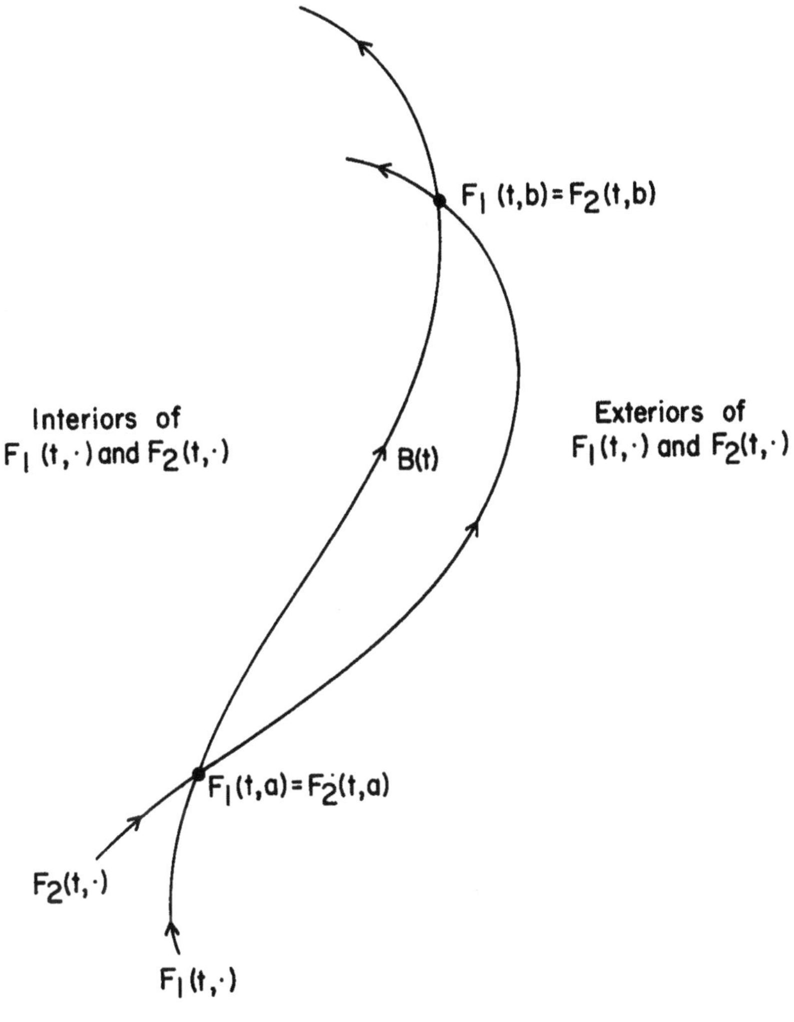

F1 (t,b) = F2 (t,b)
Interiors of
F1 (t,·) and F2 (t,·)
B(t)
Exteriors of
F1 (t,·) and F2 (t,·)
F1 (t,a) = F2 (t,a)
F2 (t,·)
F1 (t,·)

Fig. 1

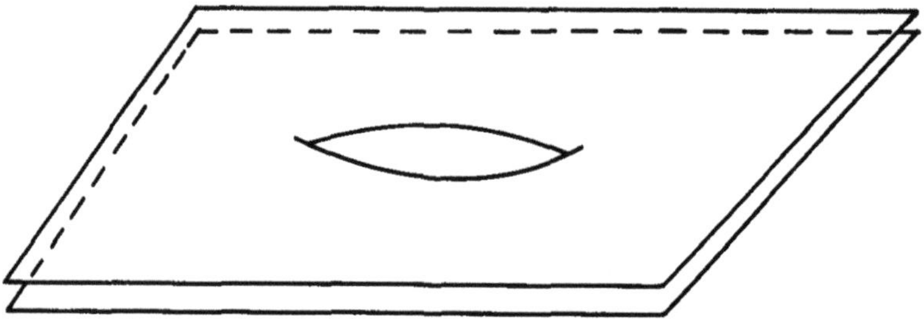

Fig. 2

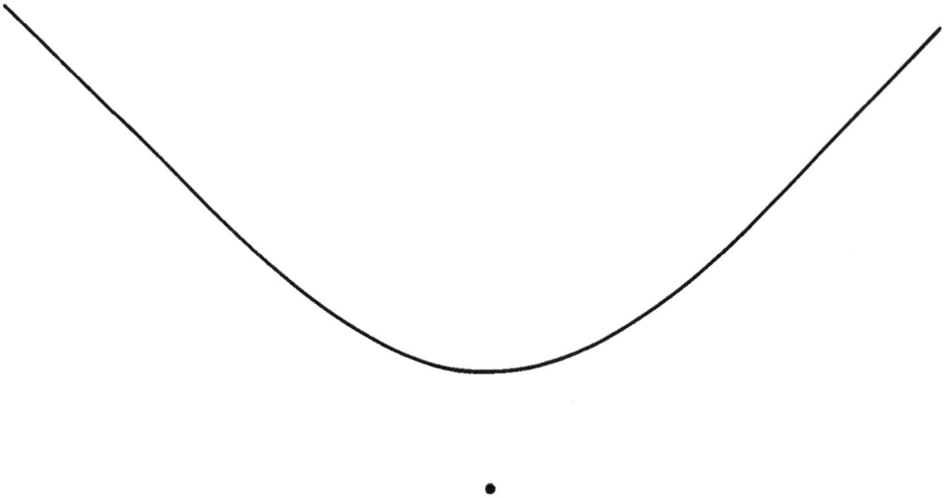

Fig. 3

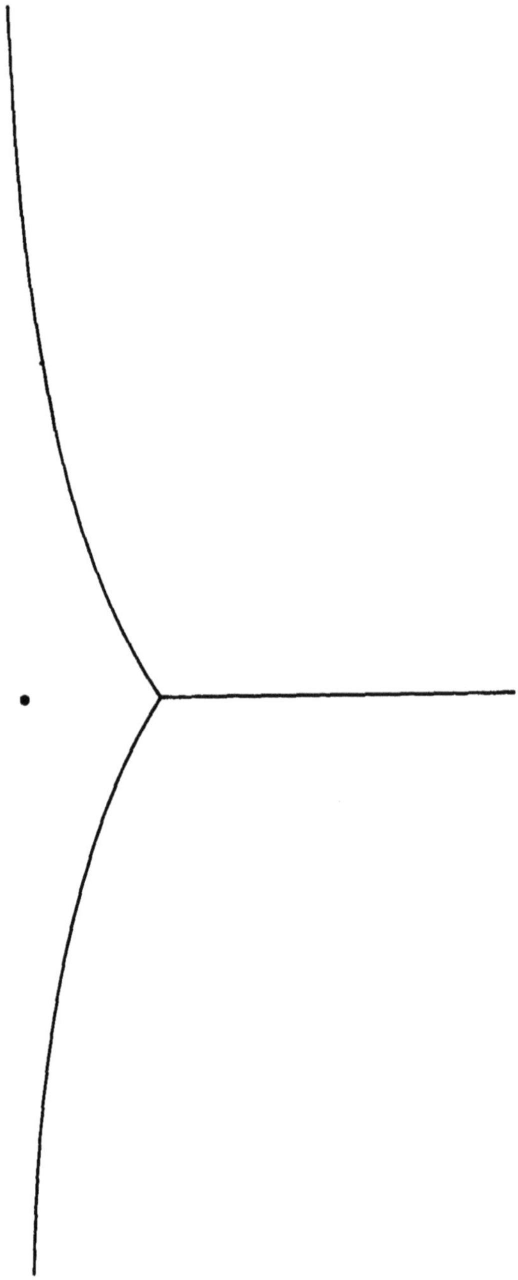

Fig. 4

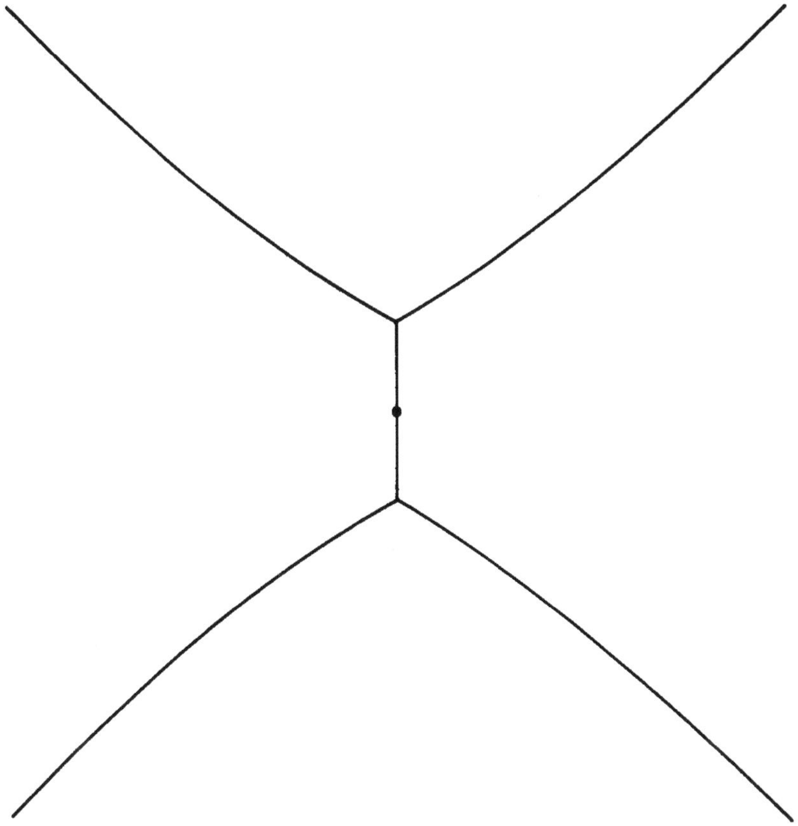

Fig. 5

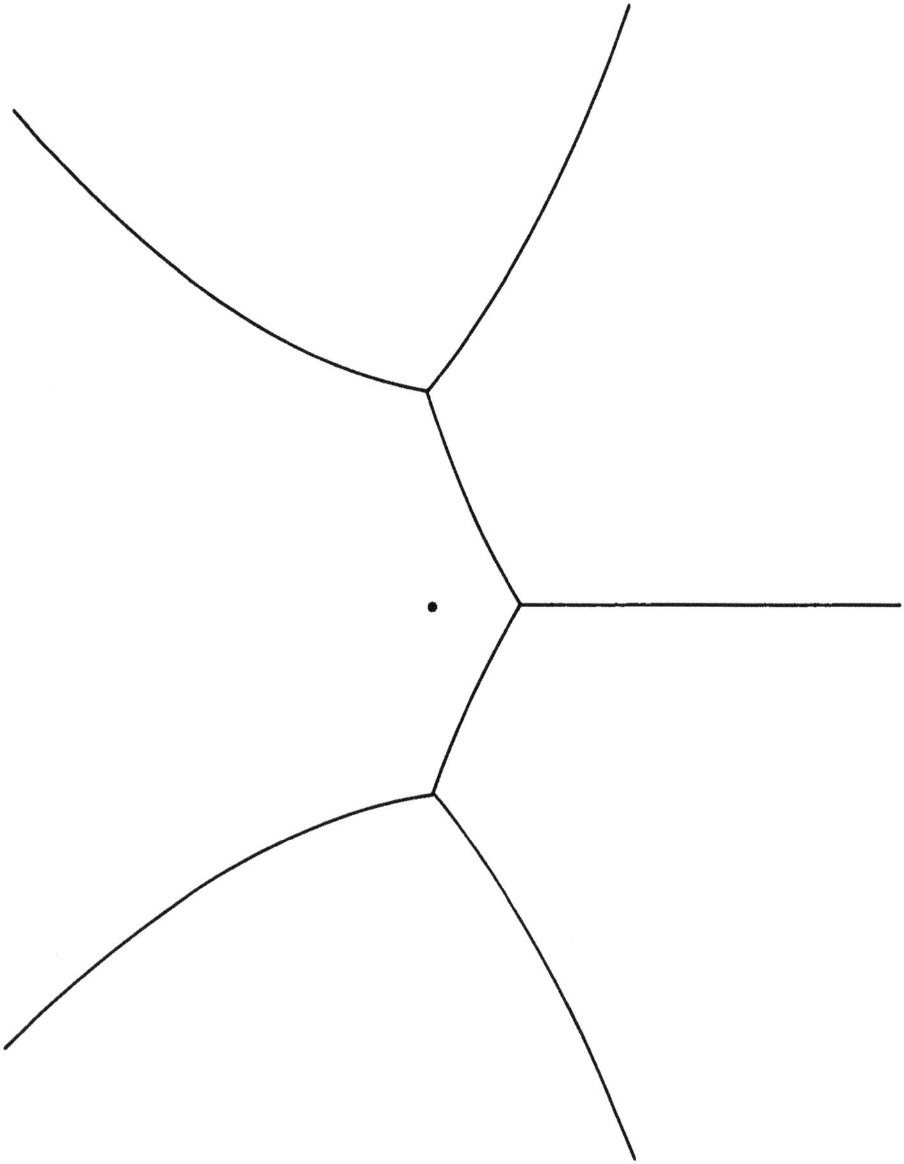

Fig. 6

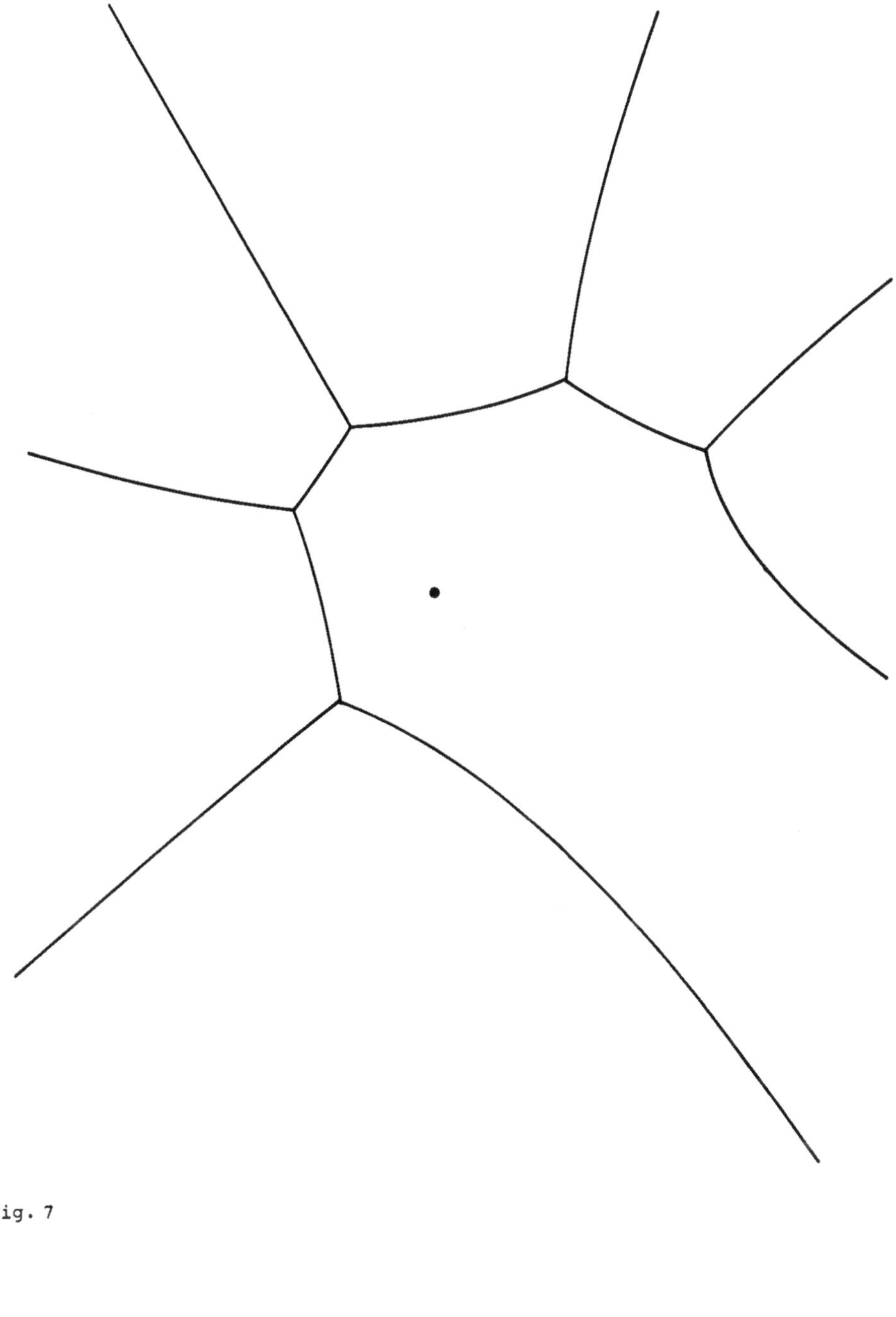

Fig. 7

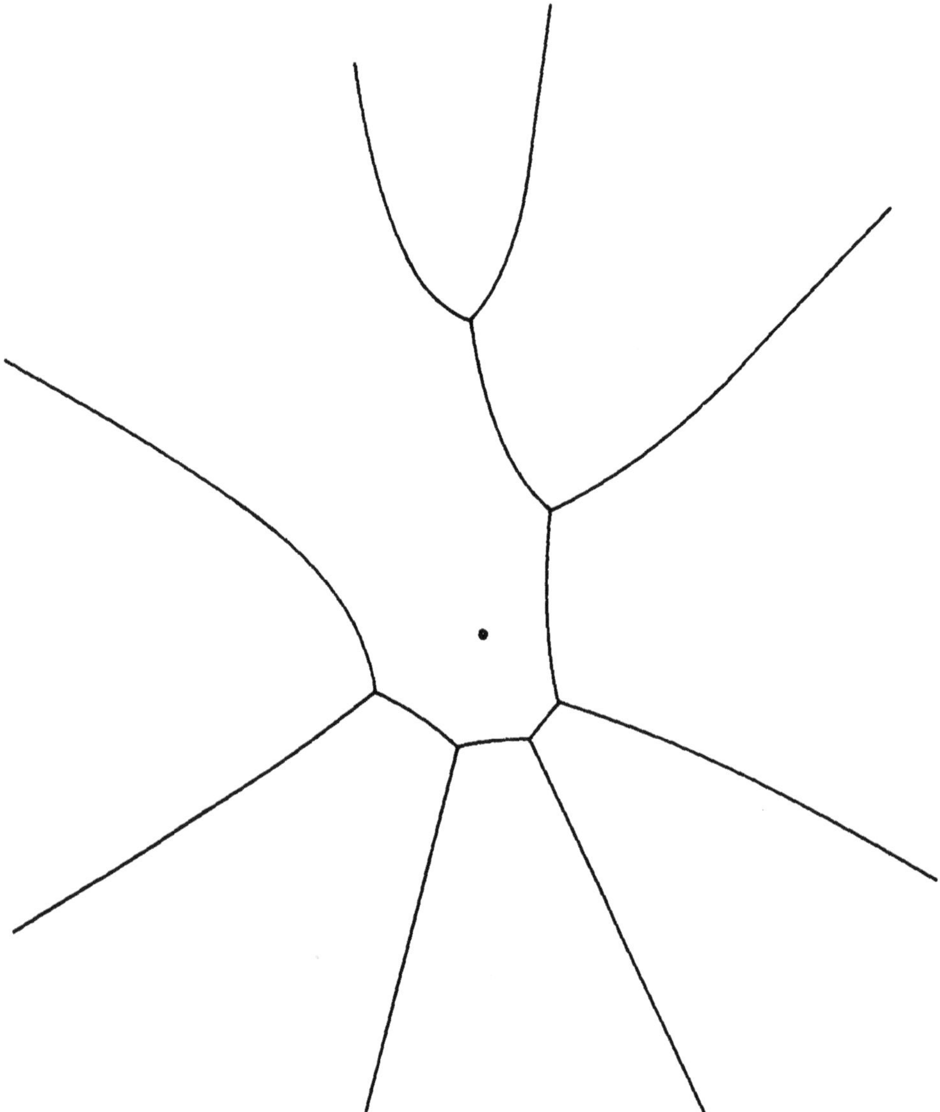

Fig. 8

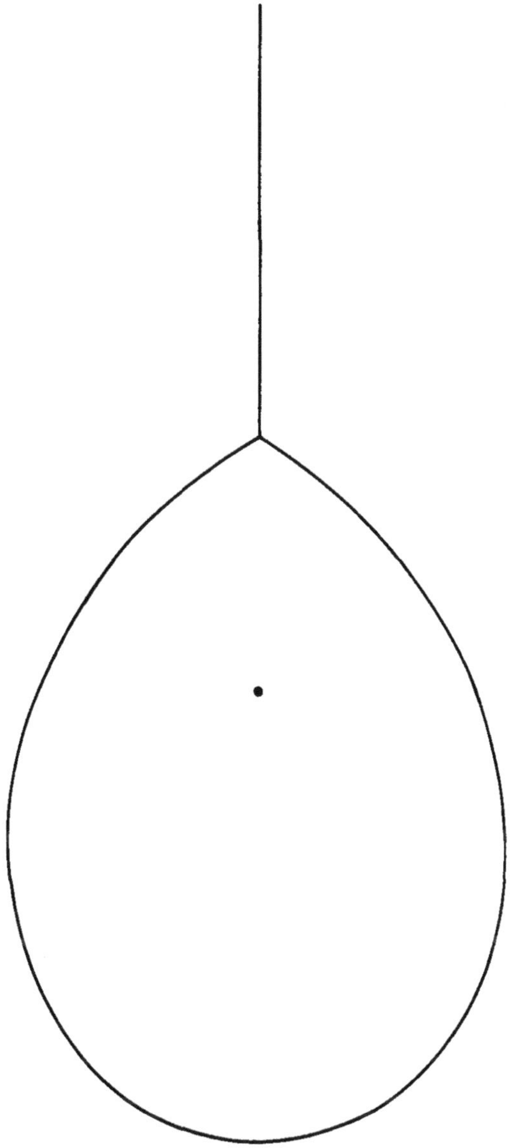

Fig. 9

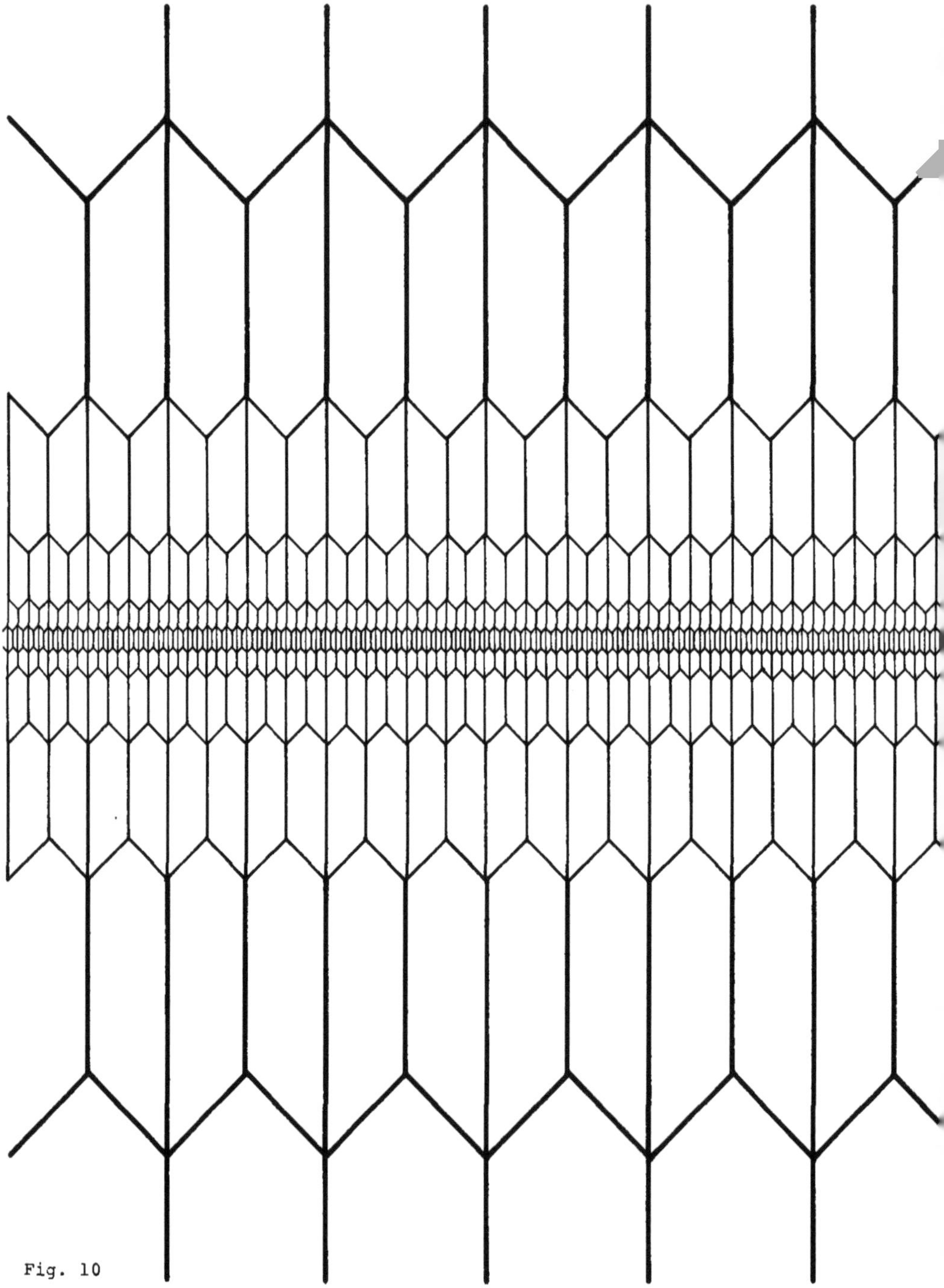

Fig. 10

REFERENCES

[AW1] W. K. Allard, On the first variation of a varifold, Ann. of Math. 95 (1972), 417-491.

[AW2] W. K. Allard, A characterization of the area integrand, Symposia Mathematica, Vol. XIV (Convegno di Teoria Geometrica dell'Integrazione e Vairetà Minimale, INDAM, Roma, Maggio 1973), pp. 429-444. Academic Press, London, 1974.

[AF] F. J. Almgren, Jr., Existence and regularity almost everywhere of solutions to elliptic variational problems with constraints, Mem. Amer. Math. Soc. no. 165 (1976).

[ES] S. D. Eidel'man, Parabolic Systems, North-Holland, Amsterdam, 1969.

[FH] H. Federer, Geometric Measure Theory, Springer-Verlag, New York, 1969.

[HCP] Handbook of Chemistry and Physics, 51st ed., Chemical Rubber Co., Cleveland, 1970.

[HC] C. Herring, Surface tension as a motivation for sintering, The Physics of Powder Metallurgy, ed. W. E. Kingston, McGraw-Hill, New York, 1951.

[K] Lord Kelvin, On homogeneous division of space with minimum partitional area, Philosophical Magazine 24 (1887), 503-514; Collected Papers, Vol. 5, p. 333.

[LT] A. Lichnewsky, R. Temam, Surfaces minimales d'evolution: le concept de pseudo-solution, C. R. Acad. Sc. Paris 284 (1977), 853-856.

[RCD] F. Rhines, K. Craig, R. Dehoff, Mechanism of steady-state grain growth in aluminum, Met. Trans. 5 (1974), 413-425.

[SC1] C. S. Smith, Metals Technol. 15 (1948).

[SC2] C. S. Smith, Grain shapes and other metallurgical applications of topology, Metal Interfaces, American Society for Metals, Cleveland, 1952.

[SM] M. Spivak, A Comprehensive Introduction to Differential Geometry, Publish or Perish, Boston, 1975.

[TJ] J. Taylor, The structure of singularities in soap bubble-like and soap film-like minimal surfaces, Ann. of Math. 103 (1976), 489-539.

Library of Congress Cataloging in Publication Data

Brakke, Kenneth A.
The motion of a surface by its mean curvature.

(Mathematical notes ; 20)
Includes bibliographical references.
1. Geometric measure theory. 2. Surfaces.
3. Curvature. I. Title. II. Series: Mathematical notes (Princeton, N.J.) ; 19.
QA312.B65 516'.1 77-19019
ISBN 0-691-08204-9

Lightning Source UK Ltd.
Milton Keynes UK
UKHW020556210822
407557UK00006B/1188

9 780691 611518